High School Algebra Essentials

Understanding Equations and Inequalities

Copyright © 2024 by Matt Maetriac
All rights reserved. No part of this publication may be reproduced, distributed, or transmitted in any form or by any means, including photocopying, recording, or other electronic or mechanical methods, without the prior written permission of the publisher, except in the case of brief quotations embodied in critical reviews and certain other noncommercial uses permitted by copyright law.

Contents

1 Foundations of Algebra   **13**
- 1.1 Understanding Variables and Constants . . 14
- 1.2 Basic Mathematical Operations 16
- 1.3 The Order of Operations (PEMDAS) 19
- 1.4 Properties of Real Numbers 21
- 1.5 Understanding and Using the Distributive Property 24
- 1.6 Combining Like Terms 27
- 1.7 Introduction to Algebraic Expressions . . . 29
- 1.8 Translating Words into Algebraic Expressions 31
- 1.9 Evaluation of Algebraic Expressions 34
- 1.10 The Concept of Sets and Subsets 36
- 1.11 Introduction to Absolute Value 39
- 1.12 Basic Principles of Equality and Inequality   42

2 Solving Linear Equations   **47**
- 2.1 Introduction to Linear Equations 47

2.2	Solving Simple Linear Equations	50
2.3	Solving Multi-Step Linear Equations	52
2.4	Variables on Both Sides of the Equation	56
2.5	Formulas and Literal Equations	59
2.6	Solving Linear Equations with Fractions	61
2.7	Applications and Word Problems	64
2.8	Solving Equations Involving Absolute Value	66
2.9	Problems Involving Proportions and Ratios	69
2.10	Introduction to Linear Inequalities	72
2.11	Solving Linear Inequalities	74
2.12	Graphical Representations of Solutions	76

3 Solving Inequalities — 81

3.1	Introduction to Inequalities	81
3.2	Solving Basic Inequalities	83
3.3	Solving Multi-Step Inequalities	86
3.4	Compound Inequalities	88
3.5	Absolute Value Inequalities	92
3.6	Inequalities Involving Quadratics	94
3.7	Rational Inequalities	98
3.8	Inequalities with Variables on Both Sides	100
3.9	Graphical Solutions of Inequalities	103
3.10	Systems of Inequalities	106
3.11	Applications of Inequalities in Word Problems	109

CONTENTS

    3.12 Introduction to Linear Programming 112

4 Graphing Linear Equations and Inequalities 115
    4.1 The Coordinate Plane 116
    4.2 Graphing Linear Equations in Two Variables 118
    4.3 Slope of a Line 120
    4.4 Forms of Linear Equations 123
    4.5 Slope-Intercept Form 125
    4.6 Point-Slope Form 128
    4.7 Standard Form of a Linear Equation 131
    4.8 Graphing Linear Inequalities in Two Variables 134
    4.9 Graphing Systems of Linear Equations . . . 137
    4.10 Graphing Systems of Linear Inequalities . . 139
    4.11 Applications of Graphing Linear Equations and Inequalities 142
    4.12 Introduction to Non-Linear Graphs 145

5 Systems of Linear Equations and Inequalities 149
    5.1 Introduction to Systems of Linear Equations 150
    5.2 Solving Systems of Equations by Graphing 152
    5.3 Solving Systems of Equations by Substitution 155
    5.4 Solving Systems of Equations by Elimination 158
    5.5 Applications and Word Problems 161
    5.6 Systems of Linear Inequalities 164
    5.7 Graphing Systems of Linear Inequalities . . 167

- 5.8 Solving Systems of Equations in Three Variables 171
- 5.9 Applications of Systems of Linear Equations and Inequalities 173
- 5.10 Introduction to Matrices 178
- 5.11 Solving Systems of Equations Using Matrices 181
- 5.12 Cramer's Rule 184

6 Polynomials 187

- 6.1 Introduction to Polynomials 187
- 6.2 Types of Polynomials 190
- 6.3 Adding and Subtracting Polynomials 192
- 6.4 Multiplying Polynomials 195
- 6.5 Special Products of Polynomials 198
- 6.6 Polynomial Long Division 200
- 6.7 Synthetic Division 202
- 6.8 The Remainder Theorem and Factor Theorem 204
- 6.9 Solving Polynomial Equations 207
- 6.10 Graphing Polynomial Functions 210
- 6.11 Polynomial Inequalities 212
- 6.12 Applications of Polynomials 214

7 Factoring Polynomials 217

- 7.1 Introduction to Factoring Polynomials . . . 218
- 7.2 Factoring out the Greatest Common Factor (GCF) 220

CONTENTS

	7.3	Factoring by Grouping 223
	7.4	Factoring Trinomials 225
	7.5	Special Factoring Formulas 227
	7.6	Factoring the Difference of Squares 230
	7.7	Factoring the Sum and Difference of Cubes 232
	7.8	Solving Polynomial Equations by Factoring 234
	7.9	Applications of Factoring Polynomials . . . 236
	7.10	The Rational Root Theorem 239
	7.11	Using Synthetic Division to Factor Polynomials 241
	7.12	Factoring Completely 243
8	**Quadratic Equations and Functions**	**247**
	8.1	Introduction to Quadratic Equations 248
	8.2	Solving Quadratic Equations by Factoring . 250
	8.3	Solving Quadratic Equations by Completing the Square 253
	8.4	Solving Quadratic Equations Using the Quadratic Formula 255
	8.5	The Discriminant and Nature of Roots . . . 258
	8.6	Applications of Quadratic Equations 261
	8.7	Graphing Quadratic Functions 263
	8.8	Properties of Parabolas 266
	8.9	Vertex Form of a Quadratic Function 269
	8.10	Transformations of Quadratic Functions . . 272
	8.11	Quadratic Inequalities 275

    8.12 The Quadratic Formula and the Discriminant 278

9 Radical Expressions and Equations   **281**
    9.1 Introduction to Radical Expressions 282
    9.2 Simplifying Radical Expressions 284
    9.3 Operations with Radical Expressions 287
    9.4 Rationalizing the Denominator 290
    9.5 Radical Equations 292
    9.6 Solving Radical Equations 295
    9.7 Complex Numbers 298
    9.8 Operations with Complex Numbers 301
    9.9 Graphing Radical Functions 303
    9.10 Applications of Radical Functions 306
    9.11 Solving Systems of Equations Involving Radicals 309
    9.12 Introduction to Imaginary and Complex Numbers 312

10 Rational Expressions and Equations   **315**
    10.1 Introduction to Rational Expressions 316
    10.2 Simplifying Rational Expressions 318
    10.3 Multiplication and Division of Rational Expressions 321
    10.4 Addition and Subtraction of Rational Expressions 324
    10.5 Complex Fractions 327
    10.6 Solving Rational Equations 330

CONTENTS

10.7 Applications of Rational Expressions 333

10.8 Rational Inequalities 336

10.9 Graphing Rational Functions 339

10.10 Asymptotes and Discontinuities 342

10.11 Inverse Variation 346

10.12 Applications of Rational Equations and Models 349

CONTENTS

Preface

This book, *High School Algebra Essentials: Understanding Equations and Inequalities*, is designed to serve as a comprehensive resource for students aiming to grasp the fundamental aspects of algebra. The objective of this book is to equip readers with a robust understanding of algebraic principles, focusing on equations and inequalities—a foundational cornerstone of mathematical education. The chapters have been carefully structured to address core topics essential for mastering high school algebra, with each chapter delving into specific aspects of algebra to ensure a thorough understanding.

The substance of the book spans various key topics, starting from the foundational elements of algebra, moving through solving different types of equations and inequalities, and culminating in understanding polynomial, radical, and rational expressions and equations. Each chapter progresses logically from basic concepts to more complex applications, ensuring that students can build their knowledge systematically. The book emphasizes problem-solving skills, offering numerous examples and exercises for students to practice and apply what they have learned.

This book is primarily intended for high school students who are either beginning their algebra studies or seeking to reinforce their understanding of algebraic concepts. It is also a valuable resource for educators looking for a structured and comprehensive algebra curriculum, as well as for parents who wish to support their children's learning. Furthermore, individuals preparing for college entrance exams or anyone interested in refreshing their algebra skills will find this book to be an invaluable resource.

In summary, *High School Algebra Essentials: Understanding Equations and Inequalities* aims to provide a solid foundation in algebra for a diverse audience. By presenting the material in a clear and logical manner, this book endeavors to make algebra accessible and understandable, allowing students to tackle mathematical challenges with confidence and proficiency.

Chapter 1

Foundations of Algebra

This chapter lays the groundwork for understanding algebra by introducing the basic elements and principles essential to the study of algebraic expressions and equations. It covers fundamental topics such as variables, constants, basic mathematical operations, and the order of operations (PEMDAS). Additionally, it elucidates the properties of real numbers, the distributive property, and combining like terms, as well as providing an overview of algebraic expressions, sets, subsets, and the concepts of absolute value and equality and inequality. This foundational knowledge is crucial for students to effectively approach more complex algebraic concepts and problems.

1.1 Understanding Variables and Constants

At the heart of algebra lies the concept of variables and constants, two fundamental entities that form the basis of algebraic expressions and equations. A clear understanding of these concepts is crucial for grasitating the intricacies of algebra. This section aims to elucidate the definitions, differences, and roles of variables and constants within the realm of algebra.

Variables are symbols that represent unknown values. They are called variables because the values they represent can vary. In algebraic expressions and equations, variables are typically denoted by letters of the alphabet. For example, in the equation $x + 3 = 5$, x is a variable that represents an unknown value that, when added to 3, equals 5. The process of algebra involves finding the value of variables that make an equation true.

On the other hand, *constants* are values that do not change. They are known quantities that are consistently the same in a given context. Constants can be any real number. For example, in the expression $3x + 4$, the numbers 3 and 4 are constants. They denote actual, fixed values in the expression, while x serves as the variable.

Understanding the distinction between variables and constants is vital for solving algebraic problems. Variables serve as placeholders for unknown quantities we seek to find, whereas constants specify fixed values that help define relationships between variables.

The Role of Variables and Constants in Algebraic Expressions

Algebraic expressions are combinations of variables, constants, and mathematical operations (such as addition, subtraction, multiplication, and division). For instance, $2x + 7$ is an algebraic expression where x is a variable, 2 and 7 are constants, and $+$ signifies addition. The value of this expression depends on the value assigned to x.

When analyzing algebraic expressions, it is crucial to identify the variables and constants as the first step. Understanding the role each part plays enables students to manipulate the expressions correctly, applying the relevant mathematical operations to solve equations.

Types of Variables

Variables are not limited to representing only one kind of quantity. In more complex scenarios, variables can denote different kinds of unknown values:

- *Dependent variables* change in response to other variables.

- *Independent variables* do not depend on other variables to define their value.

The concept of dependent and independent variables is extensively explored in functions and graphing, where the value of a dependent variable depends on the value assigned to an independent variable.

Practical Applications of Variables and Constants

Variables and constants are not abstract concepts limited to the confines of algebra; they have practical applications in various fields:

- In physics, variables can represent physical quantities such as distance, speed, or time, while constants could represent fixed values like the speed of light.

- In finance, variables might denote quantities like interest rates or investment amounts, with constants representing fixed percentages or fees.

Understanding variables and constants is foundational to algebra. Variables represent unknown values we aim to find, while constants provide fixed values that help to establish relationships in algebraic expressions. Together, they build the fundamental structure upon which algebraic reasoning is based.

1.2 Basic Mathematical Operations

In mathematics, particularly within the field of algebra, the understanding and application of basic mathematical operations are fundamental. These operations serve as the primary building blocks upon which more complex algebraic expressions and equations are constructed. There are four such basic operations that will be our focus: addition, subtraction, multiplication, and division.

1.2. BASIC MATHEMATICAL OPERATIONS

Addition

The operation of addition combines two or more quantities into a single total amount. It is denoted by the plus sign (+). Consider two numbers, a and b. The operation of adding b to a is represented by the equation:

$$a + b = c$$

where c is the sum of a and b. For example,

4 + 3 = 7

Addition is commutative, meaning that changing the order of the numbers does not affect the sum. This property can be stated algebraically as:

$$a + b = b + a$$

Subtraction

Subtraction is the operation that represents the removal of objects from a collection. The symbol for subtraction is the minus sign (-). When we subtract b from a, it is represented as:

$$a - b = c$$

where c is the difference between a and b. For instance,

5 - 2 = 3

Subtraction is not commutative, which means that $a - b$ does not necessarily equal $b - a$.

Multiplication

Multiplication is a shorthand for repeated addition. It is denoted by the symbols × or (·). The multiplication of a and b can be expressed as:

$$a \times b = c$$

or

$$a \cdot b = c$$

where c is the product. For example,

6 x 4 = 24

Multiplication is commutative, which means the order of the numbers does not affect the product:

$$a \cdot b = b \cdot a$$

Division

Division is essentially the operation of distributing a group of items into equal parts. It is represented by the symbols (/) or (÷). When we divide a by b (assuming $b \neq 0$), it is represented as:

$$a \div b = c$$

or

$$\frac{a}{b} = c$$

where c is the quotient. For instance,

20 / 5 = 4

Unlike multiplication, division is not commutative, meaning that $a \div b$ does not necessarily equal $b \div a$.

Each of these operations adheres to specific properties and rules that influence the way in which expressions containing multiple operations are solved. A profound understanding of these basic operations and their properties is crucial for the mastery of algebra. This foundation allows for the exploration and solving of more complex algebraic expressions and equations in subsequent chapters.

1.3 The Order of Operations (PEMDAS)

The correct application of the Order of Operations is fundamental in algebra and ensures that mathematical expressions are interpreted and evaluated consistently. The acronym PEMDAS stands for Parentheses, Exponents, Multiplication and Division (from left to right), and Addition and Subtraction (from left to right). This sequence guides the order in which operations should be performed to accurately solve an equation or evaluate an expression.

Parentheses have the highest priority in the order of operations. Expressions enclosed in parentheses must be evaluated first. If there are nested parentheses—the parentheses within parentheses—the evaluation starts from the innermost pair and proceeds outward.

Exponents are handled next. This includes operations involving powers and roots. If an expression contains both exponents and parentheses, the operations within the parentheses are performed prior to applying the exponent.

Following parentheses and exponents, **Multiplication and Division** are carried out from left to right. It is crucial to emphasize that multiplication and division have the same level of priority and are executed based on their order of appearance from left to right in the expression.

Similarly, **Addition and Subtraction** are the last operations to be performed, also proceeding from left to right. Like multiplication and division, addition and subtraction share the same priority level.

Let us illustrate these concepts with examples:

1. $3 + 4 \times 2 = 3 + 8 = 11$
2. $(3 + 4) \times 2 = 7 \times 2 = 14$
3. $4 + 3^2 - 1 = 4 + 9 - 1 = 12$
4. $(4 + 3)^2 - 1 = 7^2 - 1 = 49 - 1 = 48$
5. $20 \div 4 \times 3 = 5 \times 3 = 15$
6. $20 \div (4 \times 3) = 20 \div 12 = \dfrac{5}{3}$

These examples clearly demonstrate how altering the order in which operations are performed can lead to vastly different results. Thus, adherence to the PEMDAS rule is imperative for the accurate interpretation and solution of algebraic expressions.

Common Pitfalls: A frequent source of errors is the treatment of multiplication and division, as well as addition and subtraction, as if they had different levels of priority instead of being processed from left to right. It is equally essential to recognize that any operation enclosed in parentheses takes precedence over exponents adjacent to those parentheses.

In applying these principles, students are encouraged to proceed methodically, breaking down complex expressions into simpler parts and sequentially applying the PEMDAS rule. This systematic approach not only ensures accuracy but also builds a deeper understanding of the structure and logic underlying algebraic operations.

```
Example Calculations:

1. For the expression (3+4)x2, according to PEMDAS:
   - First compute the expression within the parentheses: (3+4) = 7
   - Then multiply the result by 2: 7x2 = 14

2. For the expression 20 / 4 x 3:
   - Perform the division from left to right first: 20 / 4 = 5
   - Then multiply the result by 3: 5x3 = 15
```

By adhering to the Order of Operations, students ensure clarity and prevent ambiguity in mathematical expressions, laying a solid foundation for solving more complex algebraic problems that will be introduced in subsequent chapters of this textbook.

1.4 Properties of Real Numbers

In algebra, understanding the properties of real numbers is crucial as it forms the basis upon which more complex algebraic concepts are developed. Real numbers encompass all the numbers that can be found on the number line, including whole numbers, integers, fractions, and irrational numbers. This section delves into the fundamental properties of real numbers, which are the commutative, associative, distributive, identity, and inverse properties. These properties facilitate the simplification of algebraic expressions and the solving of equations.

Commutative Property

The commutative properties of addition and multiplication state that the order in which two numbers are added or multiplied does not change the result.

For any real numbers a and b,
$a + b = b + a$ (Commutative Property of Addition)
$a \cdot b = b \cdot a$ (Commutative Property of Multiplication)

Associative Property

The associative properties of addition and multiplication assert that when three or more numbers are added or multiplied, the grouping of the numbers does not affect the sum or product.

For any real numbers $a, b,$ and c,
$(a + b) + c = a + (b + c)$ (Associative Property of Addition)
$(a \cdot b) \cdot c = a \cdot (b \cdot c)$ (Associative Property of Multiplication)

Distributive Property

The distributive property combines addition and multiplication, illustrating how a number can be distributed

1.4. PROPERTIES OF REAL NUMBERS

across a sum within parentheses to simplify an expression.

For any real numbers $a, b,$ and $c, a \cdot (b + c) = a \cdot b + a \cdot c$

Identity Property

The identity properties of addition and multiplication identify the unique number in each operation that, when combined with another number, yields the same original number.

For any real number a,
$$a + 0 = a \quad \text{(Additive Identity)}$$
$$a \cdot 1 = a \quad \text{(Multiplicative Identity)}$$

The additive identity is 0 because adding 0 to any number does not change its value. Similarly, the multiplicative identity is 1 because multiplying any number by 1 does not alter its value.

Inverse Property

The inverse properties of addition and multiplication refer to pairs of numbers that, when combined through their respective operations, yield the identity element of that operation.

For any real number a,
$$a + (-a) = 0 \quad \text{(Additive Inverse)}$$
$$a \cdot \left(\frac{1}{a}\right) = 1 \quad \text{(Multiplicative Inverse), for } a \neq 0$$

The additive inverse of a number is its negation, which, when added to the original number, results in zero. The multiplicative inverse (or reciprocal) of a number is one divided by that number, which, when multiplied by the original number, results in one.

Understanding these properties is not just an academic exercise but a practical tool that allows mathematicians and students alike to manipulate and simplify algebraic expressions and solve for unknown values efficiently. The properties of real numbers underpin many of the rules and methods used throughout algebra, providing a foundation for mathematical reasoning and problem-solving.

1.5 Understanding and Using the Distributive Property

The distributive property is a cornerstone in the field of algebra, facilitating the manipulation and simplification of algebraic expressions. It is pivotal for students to grasp this concept thoroughly, as it not only aids in simplification but also in solving equations and inequalities. This section outlines the distributive property, provides examples and exercises, and demonstrates its application in various contexts.

1.5. UNDERSTANDING AND USING THE DISTRIBUTIVE PROPERTY

The distributive property states that for any real numbers a, b, and c, the operation of multiplication over addition is performed as follows:

$$a(b + c) = ab + ac.$$

Similarly, the property holds for multiplication over subtraction:

$$a(b - c) = ab - ac.$$

The essence of the distributive property lies in its ability to distribute the operation of multiplication across the terms within the parentheses. This property is not only applicable to numbers but extends to the multiplication of variables and the combination of numbers and variables.

Examples:

Consider the algebraic expression $3(x + 4)$. Using the distributive property, we can simplify this expression as follows:

```
3(x + 4) = 3x + 12.
```

In this instance, the number 3 is distributed across both x and 4, resulting in $3x + 12$.

As another example, take the expression $-5(y - 2z)$. Applying the distributive property yields:

```
-5(y - 2z) = -5y + 10z.
```

Here, -5 multiplies both y and $-2z$ (keeping in mind that multiplying two negative numbers results in a positive number), leading to $-5y + 10z$.

Applications:

The distributive property is instrumental in various aspects of algebra, including but not limited to:

- Simplifying algebraic expressions by combining like terms.

- Solving algebraic equations and inequalities.

- Expanding expressions that involve variables and constants.

For instance, in solving the equation $4(x+3) = 20$, the first step involves using the distributive property to simplify the left side of the equation, which becomes $4x + 12 = 20$. This simplified form is conducive to further manipulation and eventual solution of the equation.

Exercises:

To reinforce understanding, the following exercises are presented for practice:

1. Simplify the expression $2(a + 5b)$ using the distributive property.

2. Given the equation $7(2x - 3) = 42$, use the distributive property as part of your strategy to find the value of x.

3. Expand the expression $y(4 - 3y + 2z)$ by applying the distributive property.

Hints for Exercises:

For exercise 1, distribute 2 across $a + 5b$. For exercise 2, start by distributing 7 across $2x - 3$, then solve for x. For exercise 3, multiply y by each term inside the parentheses.

The distributive property is a fundamental principle that underpins many operations in algebra. Mastery of this property not only simplifies the manipulation of algebraic expressions but also enriches the understanding of algebra's structure and its application in solving problems. Through consistent practice and application, students can harness the full potential of the distributive property in their algebraic endeavors.

1.6 Combining Like Terms

Combining like terms is a fundamental process in algebra that simplifies expressions and makes them easier to solve. The terms in an algebraic expression are separated by addition or subtraction signs. A like term is one that has the same variable raised to the same power. Only the coefficients of these terms can be different. This section will guide you through the process of recognizing like terms and effectively combining them by using addition or subtraction.

Consider an algebraic expression $5x + 3y - 2x + 4$. This expression consists of four terms: $5x$, $3y$, $-2x$, and 4. To simplify this expression, we identify and combine terms that are alike—terms that contain the same variables raised to the same powers. Hence, $5x$ and $-2x$ are like terms because they both contain the variable x raised to the first power. On the other hand, $3y$ and 4 are not like any other term in the expression because $3y$ is the only term containing y, and 4 is a constant without any variable.

To combine like terms, add or subtract their coefficients. For the expression mentioned, combining $5x$ and $-2x$ yields $3x$. Therefore, the simplified form of the initial expression is $3x + 3y + 4$.

Example 1. Simplify the expression $4a + 7 - 3a + 2 - a$.

First, identify the like terms, which in this case are $4a$, $-3a$, and $-a$, and the constants 7 and 2. Combining the like terms results in $4a - 3a - a$, which simplifies to $0a$ or 0, and adding the constants gives $7 + 2 = 9$. Thus, the simplified expression is 9.

```
Simplified expression: 9
```

When combining like terms, pay close attention to the operations involved. The sign in front of a coefficient is part of that term, so $-3a$ means the coefficient is -3, not 3.

Example 2. Simplify the expression $2x^2 + 3x - 4 + x^2 - 2x$.

Identify like terms: $2x^2$ and x^2 are alike because both contain x^2; $3x$ and $-2x$ are like terms because both contain x^1. The constant -4 does not have like terms. Combining like terms yields $2x^2 + x^2 = 3x^2$ and $3x - 2x = x$. Therefore, the simplified form of the expression is $3x^2 + x - 4$.

```
Simplified expression: 3x^2 + x - 4
```

Key Principles:

- Identify like terms within an expression—terms that have identical variables raised to the same power.

- Combine like terms by adding or subtracting their coefficients.

- Apply addition or subtraction to the coefficients while keeping the variables and their exponents unchanged.

- In the case of constants (terms without variables), simply add or subtract them as you would with regular numbers.

In summary, combining like terms is a crucial step in simplifying algebraic expressions. It involves identifying terms with the same variables and powers, then adding or subtracting their coefficients. This process not only makes expressions more concise but also easier to work with when solving equations or further manipulating algebraic expressions.

1.7 Introduction to Algebraic Expressions

Algebraic expressions are the cornerstone of algebra. They are combinations of variables, constants, and operation symbols (such as +, −, ×, and ÷) that together represent a number or a quantity. Understanding how to work with these expressions is critical for succeeding in algebra and beyond.

A variable is a symbol, usually a letter, that stands in for an unknown value. Constants, on the other hand, are fixed values that do not change. For example, in the expression $3x + 5$, x is a variable and 3 and 5 are constants.

Algebraic expressions come in various forms and complexity levels. The simplest form is a monomial, which is a single term consisting of a variable, a constant, or a product of both. For example, $7x$, 4, and xy^2 are all monomials.

Polynomials are algebraic expressions with more than one term. They are the sum of monomials and can be classified based on the number of terms they contain:

- A binomial has two terms, such as $x + 2$.

- A trinomial has three terms, like $2x^2 - 3x + 4$.

The degree of a polynomial is determined by the highest power of the variable present. For example, the degree of $4x^3 + 2x^2 - x + 1$ is 3, due to the term $4x^3$.

Expressions can be simplified by carrying out the indicated operations and combining like terms. Like terms are terms that contain the same variables raised to the same power. For instance, $2x$ and $3x$ are like terms and can be added to give $5x$.

The distributive property is a useful tool for simplifying expressions and is expressed as $a(b + c) = ab + ac$. It allows for the multiplication of a monomial by a polynomial, by distributing the monomial across the terms of the polynomial. For example, to simplify $3(x + 2)$, apply the distributive property to get $3x + 6$.

Evaluating an algebraic expression involves substituting the variables with their respective values and performing the operations as per the established order of operations (PEMDAS, which stands for Parentheses, Exponents, Multiplication and Division, and Addition and Subtraction). Consider the expression $2x^2 + 3x - 5$, and let $x = 2$. Substituting 2 for x gives $2(2)^2 + 3(2) - 5 = 8 + 6 - 5 = 9$.

Understanding algebraic expressions extends beyond mere manipulation and simplification. It involves recognizing the structure of expressions, which can be critical in solving equations and inequalities as well as in applications within and outside mathematics. For instance, in physics, the formula for the kinetic energy of an object, $\frac{1}{2}mv^2$, where m represents mass and v velocity, is an algebraic expression that calculates the kinetic energy of an object, showcasing the real-world application of algebraic expressions.

Moreover, translating real-life situations into algebraic expressions is a foundational skill for solving word problems. If a person buys x apples at 2 dollars each and has a 5 dollar coupon, the total cost can be represented by the algebraic expression $2x - 5$. This translation from a verbal description to an algebraic expression is crucial in effectively using algebra to solve problems.

In summary, algebraic expressions are essential in various aspects of mathematics and its applications. Mastery of their structure, simplification, evaluation, and application is fundamental for success in algebra and related fields. This foundational knowledge paves the way for more complex mathematical concepts and problem-solving strategies.

1.8 Translating Words into Algebraic Expressions

Translating verbal statements into algebraic expressions is a foundational skill in algebra that bridges the gap between real-world scenarios and mathematical reasoning. This process involves identifying keywords and phrases that correspond to mathematical operations and the use of variables to represent unknown quantities.

The first step in this translation process is understanding the role of variables. Variables are symbols, often letters such as x, y, or z, used to represent unknown numbers. When translating words into algebraic expressions, variables allow us to generalize specific situations.

Next, we need to recognize the mathematical operations

denoted by certain words or phrases. Here is a list of common terms associated with each operation:

- Addition: sum, more than, plus, increased by
- Subtraction: difference, less than, minus, decreased by
- Multiplication: product, times, of
- Division: quotient, divided by, per
- Equals: is, equals, results in

Now, let's exemplify the translation process with examples.

Example 1: Translate "The sum of a number and four" into an algebraic expression.

Recognizing the keyword "sum" refers to addition, and "a number" can be represented by a variable, let's use x. Thus, the algebraic expression is $x + 4$.

```
x + 4
```

Example 2: Translate "Three times a number decreased by five" into an algebraic expression.

Here, "three times" suggests multiplication, and "decreased by" implies subtraction. Letting the unknown number be x, the algebraic expression becomes $3x - 5$.

```
3x - 5
```

Each mathematical operation and corresponding linguistic cue must be memorized to effectively perform translations. Some phrases might lead to compound operations.

1.8. TRANSLATING WORDS INTO ALGEBRAIC EXPRESSIONS

For example, "the product of three and the sum of a number and four" involves both multiplication and addition.

To address this, we first translate the "sum of a number and four" to $x + 4$ and then multiply the entire result by three, leading to the expression $3(x + 4)$.

```
3(x + 4)
```

Variables can represent quantities in word problems, not just numbers. For instance, if a problem involves a known number of apples, a, and asks for an expression to represent the situation "twice the number of apples, decreased by three," the corresponding expression would be $2a - 3$.

Considerations for translating phrases into algebraic expressions also include understanding the importance of order. The phrase "the subtraction of a number from 10" identifies a specific order, leading to an algebraic expression of $10 - x$, not $x - 10$. The order matters significantly in operations of subtraction and division.

Lastly, compound expressions can involve several operations. For instance, translating "the sum of three times a number and five more than twice the same number" requires recognizing both "three times a number" ($3x$) and "five more than twice the same number" ($2x + 5$), leading to a composite expression $3x + (2x + 5)$.

```
3x + (2x + 5)
```

In summary, translating words into algebraic expressions requires an understanding of variables, the ability to identify keywords associated with mathematical operations, and the skill to apply these operations in the correct order. With practice, students can master this crucial skill, enabling them to solve more complex problems that stem from real-world situations.

1.9 Evaluation of Algebraic Expressions

The evaluation of algebraic expressions requires the substitution of numerical values in place of variables and performing the operations according to the established order of operations, often encapsulated by the PEMDAS rule (Parentheses, Exponents, Multiplication and Division from left to right, and Addition and Subtraction from left to right). This section will elaborate on the process of evaluating algebraic expressions, with an emphasis on clarity and proper application of mathematical operations.

An algebraic expression comprises variables, constants, and a finite series of operations. To evaluate such an expression for given values of its variables, one systematically replaces each variable with its corresponding value and computes the result following the specified sequence of operations.

Step 1: Substitute variable values. The first step involves substituting the numerical values provided for each of the variables in the expression. During this process, it is crucial to ensure that each instance of a variable is replaced with the given number.

Step 2: Follow the order of operations. After substituting the values, the next step is to perform the mathematical operations as dictated by the order of operations. It is indispensable to adhere to the PEMDAS guideline rigorously to avoid any miscalculations.

Let us consider a few examples to illustrate the evaluation of algebraic expressions.

1.9. EVALUATION OF ALGEBRAIC EXPRESSIONS

Example 1: Evaluate the expression 2x + 3y - 5 for x = 2 and y = -1.

$$\begin{aligned} 2x + 3y - 5 &= 2(2) + 3(-1) - 5 \\ &= 4 - 3 - 5 \\ &= -4 \end{aligned}$$

The final result after evaluating the expression for the given values is -4.

Example 2: Evaluate the expression (3a^2 - 2b)/(4c) for a = -1, b = 2, and c = -2.

$$\begin{aligned} \frac{3a^2 - 2b}{4c} &= \frac{3(-1)^2 - 2(2)}{4(-2)} \\ &= \frac{3(1) - 4}{-8} \\ &= \frac{-1}{-8} \\ &= \frac{1}{8} \end{aligned}$$

For the given values, the evaluated result of the expression is 1/8.

In the examples above, we observed a methodical approach of substituting the values for the variables and then executing the arithmetic operations in adherence to the PEMDAS rule. This systematic process is essential for accurately evaluating any algebraic expression.

It is noteworthy that some expressions might involve more complex operations such as exponents and roots, and the presence of parentheses implies that operations within them are performed first. The complexity of expressions can vary, but the fundamental approach of substitution followed by calculation according to the order of operations remains the core methodology.

When encountering fractions or division in expressions, extra care must be taken to ensure that division by zero does not occur. Always check the values of the variables to avoid undefined expressions due to division by zero.

In summary, the evaluation of algebraic expressions is a foundational skill in algebra that hinges on two pivotal steps: substitution of variables and adherence to the order of operations. Mastery of this process, coupled with careful attention to detail, paves the way for solving more intricate algebraic problems and understanding the behavior of functions. Regular practice with a variety of expressions can enhance one's proficiency in evaluation and contribute significantly to a deeper comprehension of algebra.

1.10 The Concept of Sets and Subsets

A set is a collection of distinct objects, which can be anything: numbers, people, letters, etc. These objects are called the elements or members of the set. Sets are fundamental to algebra and mathematics at large, providing a way to deal with collections of objects as a whole. Understanding sets and their related concepts is essential for further exploration in algebra, including the study of functions, relations, and various algebraic structures.

A set can be defined in two main ways: by listing its elements, known as the roster or tabular form, or by stating a property that its members must satisfy, known as the

1.10. THE CONCEPT OF SETS AND SUBSETS

rule or set-builder form. For example, the set of positive integers less than 5 can be expressed in roster form as

```
{1, 2, 3, 4}
```

or in set-builder form as

```
{x | x is a positive integer less than 5}.
```

When discussing sets, it is also important to understand the concept of subsets. A subset is a set whose elements are all contained in another set. The larger set is referred to as the "superset". Formally, if every element of set A is also an element of set B, then A is a subset of B, often denoted as $A \subseteq B$. It is worth mentioning that every set is a subset of itself, and the empty set, which contains no elements, is a subset of every set.

Consider the set $S = \{a, b, c, d\}$. Subsets of S include $\{a, b\}, \{b, c, d\}$, and $\{a, d\}$ among others. Remember, since every set is a subset of itself, S is also a subset of S, and the empty set, denoted as $\{\}$ or $\emptyset$, is also a subset of S.

Two special types of subsets are the proper subset and the power set. A proper subset of a set A is a subset of A that is not equal to A itself. This is denoted as $A \subsetneq B$ if A is a proper subset of B. The power set of any given set S is the set of all possible subsets of S, including S itself and the empty set. The power set is denoted as $P(S)$. For example, if $S = \{a, b\}$, then the power set of S is $\{\emptyset, \{a\}, \{b\}, \{a, b\}\}$.

The concept of cardinality refers to the number of elements within a set. The cardinality is denoted as $|S|$, where S is the set in question. For instance, if $S = \{1, 2, 3\}$ then $|S| = 3$. Understanding cardinality is crucial, especially when discussing finite and infinite sets. A set is finite if its cardinality is a specific natural number, and infinite if it has no such limit.

The notion of sets extends beyond numbers to linguistic elements, geometric figures, and much more, showcasing the versatility and foundational importance of sets in mathematics. Sets are used in defining functions, solving systems of equations, and studying relations and structures in higher mathematics. Therefore, grasping the concept of sets and subsets is a cornerstone in the apprenticeship of algebra and essentially serves as a bridge to more complex mathematical concepts and operations.

- A set is a collection of distinct objects.

- Elements can be defined by listing (roster form) or by a property (set-builder form).

- A subset is a set whose elements are all contained in another set.

- Every set is a subset of itself and the empty set is a subset of every set.

- A proper subset is a subset not equal to the set itself.

- The power set includes all possible subsets of a set.

- Cardinality refers to the number of elements in a set.

It is this comprehensive understanding of sets and subsets that lays the groundwork for moving into more detailed and complex algebraic and mathematical studies.

1.11 Introduction to Absolute Value

The concept of absolute value is essential in understanding algebra and its applications. This section explores the definition of absolute value, its mathematical notation, and its properties, alongside relevant examples and applications.

Absolute value can be defined as the distance of a number from zero on the number line, regardless of direction. The absolute value of a number a is denoted as $|a|$. It is crucial to note that the absolute value of any number is always non-negative. If a is a positive number or zero, $|a| = a$. Conversely, if a is a negative number, $|a| = -a$. This definition serves as the foundation for understanding the properties and applications of absolute value in algebra.

$$|a| = \begin{cases} a & \text{if } a \geq 0, \\ -a & \text{if } a < 0. \end{cases}$$

Properties of Absolute Value

The absolute value function possesses several key properties that are instrumental in solving algebraic problems. These include:

- **Non-negativity:** For any real number a, $|a| \geq 0$.

- **Identity Property:** For any real number a, $|a| = 0$ if and only if $a = 0$.

- **Multiplicativity:** For any real numbers a and b, $|ab| = |a| \cdot |b|$.

- **Triangle Inequality:** For any real numbers a and b, $|a + b| \leq |a| + |b|$.

Each of these properties plays a vital role in the simplification and solution of algebraic expressions and equations that involve absolute values.

Solving Equations Involving Absolute Values

Solving equations with absolute values requires understanding that the solution involves considering both the positive and negative scenarios of the values within the absolute value notation. For equation $|x| = a$, where a is a positive real number, there are two possible solutions: $x = a$ and $x = -a$.

$$|x| = a \Rightarrow x = \pm a$$

This principle can be extended to more complex equations involving absolute values.

Examples

Consider the equation $|3x - 4| = 2$. To solve this, we consider both scenarios where $3x - 4$ could be positive or negative.

$$3x - 4 = 2 \text{ or } 3x - 4 = -2$$

Solving these two equations gives us:

1.11. INTRODUCTION TO ABSOLUTE VALUE

$$3x - 4 = 2 \Rightarrow 3x = 6 \Rightarrow x = 2,$$
$$3x - 4 = -2 \Rightarrow 3x = 2 \Rightarrow x = \frac{2}{3}.$$

Thus, the equation $|3x - 4| = 2$ has two solutions: $x = 2$ and $x = \frac{2}{3}$.

Applications in Real Life

Absolute value finds applications in numerous real-life scenarios such as calculating distances, error margins in measurements, and financial models. For example, the difference in temperature from a reference point, the deviation of a data point from the mean in statistics, and the net change in a stock's price over time are all instances where absolute value is relevant.

The absolute value represents a fundamental concept in algebra that extends to various fields such as geometry, calculus, and applied mathematics. Its properties facilitate the solution of algebraic expressions and equations, making it an indispensable tool in the study and application of mathematics. Through the understanding of absolute value, students gain insights into the significance of distance and magnitude, enriching their mathematical knowledge and problem-solving skills.

1.12 Basic Principles of Equality and Inequality

The study of algebra involves examining the relationships between quantities, and a fundamental aspect of these relationships is expressed through equality and inequality. An equation signifies that two expressions are equal, while an inequality indicates that the expressions are not equal, revealing a relation of lesser or greater value.

Understanding Equations

At its simplest, an equation is a statement that two algebraic expressions are equal. It contains two expressions, one on each side of the equals sign (=). The primary goal when dealing with equations is to find the value of the variables that make the equation true.

$$x + 5 = 12$$

To solve the above equation, one would isolate the variable (x) on one side to find its value. This is achieved by performing the same operation on both sides of the equation, maintaining its balance.

$$x + 5 - 5 = 12 - 5$$
$$x = 7$$

The solution shows that when $x = 7$, both sides of the equation are equal, thus making the equation true. This

1.12. BASIC PRINCIPLES OF EQUALITY AND INEQUALITY

process demonstrates the principle of balancing, pivotal in solving equations.

Exploring Inequalities

In contrast to equations, inequalities express a different kind of relationship between expressions. They indicate that one expression is less than ($<$), greater than ($>$), less than or equal to ($\leq$), or greater than or equal to ($\geq$) another expression. Solving inequalities involves finding all possible values of the variables that make the inequality true.

Consider the following inequality as an example:

$$2x - 3 < 5$$

To solve for x, similar steps as solving equations are used, with careful attention to the inequality sign.

$$2x - 3 + 3 < 5 + 3$$
$$2x < 8$$
$$x < 4$$

The solution indicates that any value of x less than 4 would satisfy the inequality.

Properties of Equality and Inequality

Several properties govern the operations involving equalities and inequalities, ensuring mathematical consistency and aiding in problem-solving.

- **Addition and Subtraction Properties:** For equations, if the same amount is added or subtracted from both sides, the balance is maintained. For inequalities, doing the same does not change the direction of the inequality.

- **Multiplication and Division Properties:** In the case of equations, multiplying or dividing both sides by the same non-zero number maintains the equation's balance. For inequalities, multiplying or dividing by a positive number does not alter the inequality's direction, but doing so by a negative number reverses the inequality's direction.

- **Substitution Property:** If two algebraic expressions are equal, one can replace one with the other in any equation or inequality without affecting the truth value.

It is important to note that while solving inequalities, reversing the inequality sign when multiplying or dividing by a negative number is critical. This distinction from equations ensures the logical consistency of inequalities.

Applying Equality and Inequality

Understanding and applying the principles of equality and inequality are foundational in algebra. They not only facilitate the solving of equations and inequalities but also the analysis and understanding of mathematical relationships. Mastery of these principles enables students to approach more complex algebraic problems with confidence.

1.12. BASIC PRINCIPLES OF EQUALITY AND INEQUALITY

The essential nature of equality and inequality within algebra serves as a cornerstone for understanding mathematical relationships. By mastering these principles, one gains the ability to solve a wide array of algebraic problems, set the stage for exploring functions and their graphs, and lay the groundwork for more advanced mathematical topics.

Chapter 2

Solving Linear Equations

This chapter delves into the techniques and methodologies necessary for solving linear equations, which are equations of the first degree. It begins with simplifying and solving basic linear equations and progresses to handling multi-step equations, equations with variables on both sides, and equations involving absolute values. The chapter also discusses solving linear equations with fractions, applications in word problems, and introduces linear inequalities. Emphasis is placed on developing problem-solving skills through a systematic approach, helping students to understand not just how to find solutions but also why these methods are effective.

2.1 Introduction to Linear Equations

A linear equation in one variable is an equation that can be expressed in the form $ax + b = 0$, where a and b are real numbers and a is not equal to zero. The term "linear" indicates that the graph of the equation, in a two-dimensional coordinate system, forms a straight line. The solutions to these equations are the values of x that satisfy the equation, making the expression true.

The simplicity of linear equations makes them a fundamental concept in algebra and a starting point for understanding more complex mathematical relationships. The ability to solve these equations is not only a key skill for algebra students but also a crucial tool in various applied fields such as physics, engineering, economics, and beyond.

Properties of Linear Equations

Linear equations possess several properties that aid in their solution. These include:

- The solution to a linear equation does not change if the same number is added or subtracted from both sides of the equation.

- Multiplying or dividing both sides of a linear equation by the same non-zero number does not alter the solution.

- If a linear equation is true, any multiple of both sides of the equation is also true.

- If two linear equations are true, their sum or difference is also a true statement.

Leveraging these properties allows for the manipulation of equations into forms that are easier to solve, usually aiming to isolate the variable of interest on one side of the equation.

2.1. INTRODUCTION TO LINEAR EQUATIONS

Solving a Linear Equation

The goal in solving any linear equation is to find the value of the variable that makes the equation true. The process generally involves a few key steps:

1. Simplify both sides of the equation, if necessary, by distributing multiplications over additions and subtracting like terms.

2. If the variable appears on both sides of the equation, use addition or subtraction to get all terms with the variable on one side and constants on the other.

3. Isolate the variable by performing any necessary additions or subtractions to remove constants from the side of the equation with the variable.

4. Finally, solve for the variable by dividing both sides of the equation by the coefficient in front of the variable.

Example: Solving a Basic Linear Equation

Consider the equation $3x + 5 = 20$. To solve for x, we aim to isolate x on one side:

$$\begin{aligned} 3x + 5 &= 20 \\ 3x &= 20 - 5 \quad &&\text{(Subtract 5 from both sides)} \\ 3x &= 15 \\ x &= \frac{15}{3} \quad &&\text{(Divide both sides by 3)} \\ x &= 5 \end{aligned}$$

Thus, the solution to the equation $3x + 5 = 20$ is $x = 5$.

x = 5

This procedure demonstrates the systematic approach to solving linear equations, emphasizing the importance of performing the same operation on both sides of the equation to maintain its equality. Through practice, this methodical process becomes intuitive, allowing students to solve increasingly complex linear equations with confidence.

Linear equations are the cornerstone of algebra, bridging the gap between basic arithmetic and more advanced mathematics. Mastery of linear equations equips students with the tools necessary for tackling a wide range of mathematical problems, serving as the groundwork for exploring linear inequalities, systems of equations, and beyond.

2.2 Solving Simple Linear Equations

Linear equations are foundational to algebra and critical for understanding numerous real-world phenomena. A simple linear equation is an equation of the first degree, meaning it contains only one variable raised to the first power and can be represented in the form $ax + b = 0$, where a and b are constants. An effective approach to solving such equations is the focus of this section.

Solving a simple linear equation requires isolating the variable on one side of the equation. This process involves a series of equivalent transformations that do not change the equation's solutions. We will examine these

2.2. SOLVING SIMPLE LINEAR EQUATIONS

steps systematically, ensuring a solid grasp of each operation.

Step 1: Simplify both sides of the equation, if necessary. This includes expanding any parentheses and combining like terms. For instance, if we start with $2(3x - 4) + x = 5x - 8$, simplification yields $6x - 8 + x = 5x - 8$, and further reduces to $7x - 8 = 5x - 8$.

Step 2: Move all the variable terms to one side of the equation. This is achieved by adding or subtracting terms on both sides. Continuing with our example, subtract $5x$ from both sides to get $2x - 8 = -8$.

Step 3: Isolate the variable. Once the variable is on one side, isolate it by performing operations that undo the addition or multiplication involving the variable. In our case, add 8 to both sides to eliminate the constant term next to the variable, resulting in $2x = 0$. Dividing both sides by 2 gives us $x = 0$.

$$2x = 0 \Rightarrow x = 0$$

This solution suggests that the original equation holds true when x equals zero. The steps outlined above can be succinctly summarized and applied to solve any simple linear equation.

Examples

Consider solving the equation $3x - 9 = 0$.

Applying Step 1, the equation is already simplified. Moving to Step 2 and 3, adding 9 to both sides gives $3x = 9$, and dividing both sides by 3 yields the solution:

x = 3

As another example, solve $-4x + 2 = 10$.

First, we subtract 2 from both sides, resulting in $-4x = 8$. Dividing both sides by -4 gives:

x = -2

The simplicity of linear equations allows for the systematic application of algebraic principles to uncover the value of the unknown variable. The strength of these methods is their universality and their ability to simplify and solve equations that might initially seem complex.

Practice Problems

1. Solve the equation $5x - 15 = 0$.

2. Solve $7 - 2x = 3x - 8$.

3. Determine the solution to $4(2x - 1) = 8$.

Solutions to these problems reinforce the steps of simplifying the equation, moving variables to one side, and performing algebraic operations to isolate the variable. Solving simple linear equations is not just about finding answers but understanding the rationale behind each operation and its role in the overall process of equation solving. Mastery of these techniques forms the bedrock for exploring more complex algebraic concepts.

2.3 Solving Multi-Step Linear Equations

Solving multi-step linear equations is a process that involves performing a series of operations to isolate the variable and find its value. These types of equations often require more than one algebraic step to solve. This section outlines a structured approach to tackling these equa-

2.3. SOLVING MULTI-STEP LINEAR EQUATIONS

tions, providing clear examples and common strategies to guide students through the process.

First, let's define a multi-step linear equation: it is an equation that requires more than one operation, such as addition, subtraction, multiplication, or division, along with the possible simplification of terms, to solve for the unknown variable.

Key Steps in Solving Multi-Step Linear Equations

The general strategy to solve multi-step linear equations involves the following steps:

- Simplify both sides of the equation, if necessary, by removing parentheses and combining like terms.
- Move all terms containing the variable to one side of the equation and constants to the other.
- Combine like terms on each side of the equation.
- Isolate the variable by performing operations that will leave the variable on one side of the equation and a numerical value on the other.
- Check the solution by substituting it back into the original equation to verify that the equation is true.

Example

Consider the equation $2(3x - 4) - 5(x - 2) = 3(x + 2) - 7$.

Step 1: Simplify both sides of the equation.

$$2(3x - 4) - 5(x - 2) = 3(x + 2) - 7$$
$$6x - 8 - 5x + 10 = 3x + 6 - 7$$
$$x + 2 = 3x - 1$$

Step 2: Move all terms containing the variable to one side and all constants to the other.

$$x - 3x = -1 - 2$$
$$-2x = -3$$

Step 3: Solve for the variable.

$$x = \frac{-3}{-2}$$
$$x = \frac{3}{2}$$

Step 4: Check the solution.

Substitute $x = \frac{3}{2}$ into the original equation.

$$2\left(3\left(\frac{3}{2}\right) - 4\right) - 5\left(\left(\frac{3}{2}\right) - 2\right) = 3\left(\left(\frac{3}{2}\right) + 2\right) - 7$$
$$6 - 8 - 5(-\frac{1}{2}) = 4.5 + 6 - 7$$
$$-2 + \frac{5}{2} = \frac{9}{2} - 7$$
$$\frac{1}{2} = \frac{1}{2}$$

The solution $x = \frac{3}{2}$ satisfies the original equation, thus verifying our solution.

2.3. SOLVING MULTI-STEP LINEAR EQUATIONS

Common Mistakes and Tips

When solving multi-step linear equations, it is essential to avoid common pitfalls:

- Failing to simplify the equation fully before attempting to solve for the variable.
- Neglecting to distribute multiplication over addition or subtraction in parentheses properly.
- Forgetting to perform the same operation on both sides of the equation to maintain equality.
- Combining unlike terms mistakenly.

To navigate these complex equations successfully, students are encouraged to:

- Write each step clearly and neatly to avoid confusion.
- Check work at each stage to catch and correct any errors promptly.
- Practice with various equations to become familiar with different types of problems.
- Seek assistance when struggling with particular steps or operations.

Solving multi-step linear equations is a fundamental skill in algebra that provides a foundation for solving more complex problems. With practice and careful attention to the outlined steps and common pitfalls, students can enhance their problem-solving capabilities in this critical area of mathematics.

2.4 Variables on Both Sides of the Equation

In many algebraic equations, variables appear on both sides of the equals sign. This situation requires a careful, systematic approach to isolate the variable and find its value. This section explores the strategies and steps involved in solving equations that have variables on both sides. Understanding this concept is crucial, as it broadens your problem-solving toolkit and prepares you for more complex algebraic challenges.

When faced with an equation that has variables on both sides, the key objective is to manipulate the equation in such a way that all the variable terms are consolidated on one side, and all the constant terms are on the other. This process involves three major steps: distributing (if necessary), combining like terms, and then isolating the variable.

Step 1: Distribution

Often, equations with variables on both sides include parentheses that need to be dealt with first. This is accomplished through the distributive property $a(b + c) = ab + ac$. This step ensures that any grouped terms are expanded, making it easier to identify and move like terms in subsequent steps.

Step 2: Combining Like Terms

Once all parentheses are eliminated, the next step is to combine like terms on each side of the equation. Like terms are terms that have the exact same variable raised to the same power. Combining them simplifies the equation and reduces the complexity of the problem.

Step 3: Isolating the Variable

This step focuses on moving all the variable terms to one side and all the constant terms to the other. This is achieved by adding or subtracting terms on both sides of the equation. The goal is to have all terms involving the variable on one side of the equation, and all the numerical (constant) terms on the other side.

Once the variables are on one side, the equation may sometimes still need to be simplified by combining like terms again or by dividing both sides of the equation to isolate the variable completely.

Example:

Consider the equation $3x + 4 = 7x - 8$.

1. First, subtract $3x$ from both sides to get the variables on one side:
$$3x + 4 - 3x = 7x - 8 - 3x$$
$$4 = 4x - 8$$

2. Next, add 8 to both sides to get the constant terms on the opposite side:
$$4 + 8 = 4x - 8 + 8$$
$$12 = 4x$$

3. Finally, divide both sides by 4 to isolate the variable:
$$\frac{12}{4} = \frac{4x}{4}$$
$$3 = x$$

Solution: x = 3

In this example, we systematically manipulated the equation by first dealing with the distribution step (which wasn't necessary in this case), followed by combining like terms and isolating the variable. This process involved addition, subtraction, and division to reach the solution.

Practice Problems:

To master solving equations with variables on both sides, attempt the following practice problems.

- $2x + 3 = 3x - 2$
- $5(y + 2) = 2y - (3 - 4y)$
- $4a + 5 - 3a = 2(2a - 1) + 3$

It is essential to work through these problems methodically, applying the steps outlined: distribution if necessary, combining like terms, and isolating the variable. Through consistent practice, solving equations with variables on both sides will become a more manageable and intuitive process.

In summary, equations with variables on both sides pose an interesting challenge but are surmountable with a clear, step-by-step approach. By distributing, combining like terms, and isolating variables, you can simplify and solve these equations effectively. This foundational skill sets the stage for tackling more advanced algebraic concepts and applications.

2.5 Formulas and Literal Equations

The mastery of handling formulas and solving for literal equations is a cornerstone skill in algebra, finding applications across mathematics, physics, engineering, and economics. This section focuses on manipulating algebraic formulas, a process crucial for deriving specific variable values that these formulas relate. Unlike the previous sections where the objective was to solve for a numerical value, here we deal with equations involving multiple variables, aiming to solve for one variable in terms of others.

Understanding the principles of algebraic manipulation is fundamental. The techniques of adding, subtracting, multiplying, and dividing both sides of an equation by the same value, discussed in earlier sections, are directly applicable. The primary difference lies in the goal: isolating a variable rather than finding its specific value.

General Approach: The process typically begins by identifying the variable to solve for, then systematically performing operations to isolate this variable on one side of the equation. It is essential to maintain the balance of the equation by uniformly applying operations across both sides.

Example 1: Solve for r in the formula for the area of a circle, $A = r^2$.

Step 1: Divide both sides by to get $A/ = r^2$.
Step 2: Apply the square root to both sides to isolate r, giving $r = sqrt(A/)$.

This example illustrates the step-by-step manipulation of a formula to isolate a desired variable.

Working with Literal Equations: Literal equations, unlike standard linear equations, may not be solved for a numerical answer but rather manipulated to solve for one variable in terms of others. This requires a deep understanding of algebraic properties and the ability to apply them in various situations.

Example 2: Solve for x in terms of a, b, and c from the equation ax + b = c.

To solve for x, we perform operations to isolate x on one side of the equation:

$$ax + b = c$$
$$\text{Subtract } b \text{ from both sides:} \quad ax = c - b$$
$$\text{Divide both sides by } a: \quad x = \frac{c - b}{a}$$

Complex Manipulations: Some formulas require more complex manipulations, involving distributing, factoring, or working with fractions. Consideration of the order of operations is paramount in such scenarios.

```
Example 3: Solve for y in the equation 3/(2y) + 5 = 9.

Step 1: Subtract 5 from both sides to get 3/(2y) = 4.
Step 2: Multiply both sides by 2y to eliminate the fraction, leading to
 3 = 8y.
Step 3: Divide both sides by 8 to isolate y, yielding y = 3/8.
```

Application in Real-World Problems: The ability to manipulate and rearrange formulas is not merely an academic exercise but a practical skill. In physics, for instance, the equation for force (F = ma, where F is force, m is mass, and a is acceleration) can be rearranged to solve for mass or acceleration depending on the given variables and the desired unknown. In finance, solving for the interest rate, principal, or period in the formula for compound interest necessitates similar algebraic manipulation.

- Understanding and identifying the role of each variable within an equation is critical.
- Step-by-step isolation of the desired variable, ensuring each operation is legally applied across the equation to maintain its balance, is necessary.
- Recognizing when to apply specific algebraic techniques, such as distributing or combining like terms, is part of the skill set required.

In summary, navigating through formulas and literal equations demands a solid grasp of algebraic principles and the ability to apply them across a spectrum of variables and equations. This proficiency not only lays the foundation for advanced mathematical problem solving but also enhances logical reasoning and analytical skills essential in various professional fields.

2.6 Solving Linear Equations with Fractions

Linear equations with fractions may initially seem more challenging due to the presence of denominators. However, by applying foundational algebraic principles, these equations can be solved with precision and efficiency. The key strategy involves eliminating the fractions at the earliest stage by finding a common denominator for all terms in the equation. This section will elucidate the methodical steps required to simplify and solve linear equations that involve fractions.

To begin, consider a general linear equation with fractions. The equation can be represented as $\frac{a}{b}x + \frac{c}{d} = \frac{e}{f}$, where a, b, c, d, e, and f are constants, and x is the variable we aim to solve for. The goal is to isolate x on one side of the equation.

The first step in solving equations with fractions is to find a common denominator among all the fractions present. This common denominator, often denoted as LCD (Least Common Denominator), allows us to rewrite each term as an equivalent fraction with the same denominator, thereby facilitating the simplification process.

$$LCD = b \cdot d \cdot f$$

Once the LCD is determined, both sides of the equation are multiplied by it. This action effectively removes the denominators, simplifying the equation to one that involves only integers and the variable.

$$LCD \cdot \frac{a}{b}x + LCD \cdot \frac{c}{d} = LCD \cdot \frac{e}{f}$$
$$b \cdot d \cdot f \cdot \frac{a}{b}x + b \cdot d \cdot f \cdot \frac{c}{d} = b \cdot d \cdot f \cdot \frac{e}{f}$$

Simplifying each term by cancelling out common factors leads to a more straightforward equation:

$$a \cdot d \cdot f \cdot x + c \cdot b \cdot f = e \cdot b \cdot d$$

At this juncture, the equation no longer contains fractions and can be approached as any standard linear equation.

2.6. SOLVING LINEAR EQUATIONS WITH FRACTIONS

Applying principles of algebra, one can isolate x by performing suitable arithmetic operations.

Let us illustrate the outlined procedure with an example:

```
Example: Solve the equation x - ½ = ¼x + .
```

First, identify the LCD of the denominators 3, 2, 4, and 5. In this case, $LCD = 60$.

$$60 \cdot \left(\frac{1}{3}x - \frac{1}{2}\right) = 60 \cdot \left(\frac{1}{4}x + \frac{1}{5}\right)$$
$$20x - 30 = 15x + 12$$

Next, isolate x:

$$20x - 15x = 12 + 30$$
$$5x = 42$$
$$x = \frac{42}{5}$$

The solution to the equation is $x = \frac{42}{5}$.

```
Result: x = 8.4
```

This example demonstrates the importance of eliminating fractions early in the process by finding and using the LCD. Through this methodical approach, linear equations with fractions can be resolved to reveal precise solutions.

In summary, solving linear equations with fractions necessitates an understanding of how to find a common denominator and use it to eliminate the fractions from the equation. This approach, combined with algebraic manipulation to isolate the variable, paves the way for solving such equations systematically and effectively. This section has aimed to equip students with the necessary tools and confidence to tackle linear equations with fractions, thereby broadening their problem-solving repertoire.

2.7 Applications and Word Problems

Linear equations are not just abstract concepts confined to the classroom; they have extensive applications in real-world problems. These applications range from calculating distances, times, and rates to financial planning and beyond. This section explores how to translate real-life situations into linear equations and solve them to find practical solutions. Mastery of these concepts enables students to apply their mathematical skills outside of an academic context, providing them with tools to solve a variety of practical problems.

To approach word problems effectively, one must first understand how to translate the given information into a mathematical form. This often involves identifying quantities and their relationships as described in the problem, then expressing these relationships as linear equations.

Consider a basic framework for solving word problems:

- Read the problem carefully, ensuring that you understand all given information and what is being asked.
- Identify and assign variables to unknown quantities.

2.7. APPLICATIONS AND WORD PROBLEMS

- Translate the word problem into a linear equation using the given information and relationships between the quantities.

- Solve the linear equation.

- Check the solution by inserting it back into the context of the problem.

- Clearly state the answer, including units if applicable.

Let us apply this framework to some examples to demonstrate how to translate and solve word problems involving linear equations.

Example 1: A rental car company charges a basic fee of $50 per day for renting a car and an additional charge of $0.20 per mile driven. Write a linear equation to represent the total cost, C, of renting the car for one day and driving it x miles.

Solution: Here, the total cost (C) depends on two components: a fixed charge ($50) and a variable charge that depends on the number of miles driven ($0.20x$). Thus, the linear equation can be represented as:

$$C = 50 + 0.20x$$

Example 2: A company has a monthly fixed operating cost of $5000 and produces widgets at a variable cost of $2 per widget. If x represents the number of widgets produced, and C represents the total monthly cost, formulate a linear equation relating C and x.

Solution: The total cost (C) is the sum of the fixed cost ($5000) and the variable cost ($2x$), which leads to the following equation:

$$C = 5000 + 2x$$

Example 3: Two bikers are 30 miles apart and are cycling toward each other. One biker is cycling at 10 miles per hour, and the other is cycling at 15 miles per hour. How long, in hours, will it take for the two bikers to meet?

Solution: Let t represent the time in hours it takes for the bikers to meet. In t hours, the first biker will have traveled $10t$ miles, and the second biker will have traveled $15t$ miles. Since they are cycling toward each other, the sum of these distances must equal the initial distance between them, which is 30 miles. This results in the equation:

$$10t + 15t = 30$$

Solving this equation for t yields:

```
t = 1
```

Thus, it will take 1 hour for the two bikers to meet.

Solving word problems involving linear equations requires careful reading and translation of the given problem into a mathematical model. By identifying variables, creating an equation, and solving for the unknowns, students can tackle a wide range of real-world problems. Practice with varied examples helps in developing this skill, bridging the gap between theoretical mathematics and its practical applications.

2.8 Solving Equations Involving Absolute Value

The absolute value of a number refers to its distance from zero on the number line, regardless of direction. Thus, it is always non-negative. An absolute value equation is

2.8. SOLVING EQUATIONS INVOLVING ABSOLUTE VALUE

an equation that contains an absolute value expression. Solving equations involving absolute values is considerably distinct from solving standard linear equations due to the dual nature of absolute value, representing both positive and negative outcomes. This section elucidates methods for solving such equations, emphasizing an analytical step-by-step approach.

The fundamental principle underlying the solution of absolute value equations is the definition itself. If $|x| = a$, where a is a positive number, then $x = a$ or $x = -a$.

Basic Form: An absolute value equation typically takes the form $|ax + b| = c$, where a, b, and c are constants, and c is non-negative. To solve such equations, one must consider both positive and negative scenarios that satisfy the equation.

Step-by-Step Solution:

- First, isolate the absolute value expression on one side of the equation. If the equation is $|ax + b| = c$, ensure it is in the form before proceeding.

- Next, set up two separate equations to account for the dual nature of absolute value: one for the positive scenario $(ax + b = c)$ and another for the negative scenario $(ax + b = -c)$.

- Solve each of these linear equations independently.

- Verify the solutions by plugging them back into the original equation to ensure they satisfy the absolute value condition.

Example: Solve the equation $|2x - 3| = 7$.

To solve this equation, we consider both scenarios where the expression inside the absolute value could be positive or negative but equal to 7 in magnitude.

Positive Scenario: $2x - 3 = 7$

$$2x = 7 + 3$$
$$2x = 10$$
$$x = 5$$

Negative Scenario: $2x - 3 = -7$

$$2x = -7 + 3$$
$$2x = -4$$
$$x = -2$$

Solution Verification: Plug $x = 5$ and $x = -2$ back into the original equation to verify.

```
|2(5) - 3| = 7   ->   |10 - 3| = 7   ->   |7| = 7 (True)
|2(-2) - 3| = 7  ->   |-4 - 3| = 7   ->   |-7| = 7 (True)
```

Both solutions, $x = 5$ and $x = -2$, are valid.

Special Cases:

- If the equation is in the form $|ax + b| = 0$, the only solution is when $ax + b = 0$ since absolute value can only be zero if the expression inside is zero.

- If $|ax + b| = c$, where c is a negative number, the equation has no solution. Absolute value cannot equal a negative number as it represents distance.

- Equations involving variables both inside and outside the absolute value sign require careful isolation of the absolute value expression before setting up the two scenarios.

Word Problems and Applications: Absolute value equations often appear in real-world contexts where distance and magnitude are discussed. For example, determining the possible values of a measurement error, or the distances that can be traveled in opposite directions, come down to solving absolute value equations. These applications underscore the importance of understanding both the mathematical technique and the conceptual underpinning of absolute values.

Solving equations involving absolute values is a critical skill in algebra. It requires an analytical approach that not only leverages the basic properties of absolute values but also integrates problem-solving and algebraic manipulation strategies. Mastery of this topic enhances students' ability to tackle a broad spectrum of mathematical problems, including those rooted in real-life situations.

2.9 Problems Involving Proportions and Ratios

Problems involving proportions and ratios are common in algebra and are applicable in various real-world scenarios, including but not limited to, rate problems, mixture problems, and scale models. Understanding how to solve these problems is crucial for students' mathematical development.

A ratio is a way to compare two quantities by division, expressed in the form $a : b$ or, equivalently, as the fraction $\frac{a}{b}$, where $b \neq 0$. A proportion, on the other hand, is an equation that states two ratios are equal. It can be represented as $\frac{a}{b} = \frac{c}{d}$, where $a, b, c,$ and d are real numbers and $b, d \neq 0$.

Solving Proportions

To solve a proportion for an unknown, one could employ the cross-multiplication technique, which involves multiplying the denominator of each ratio by the numerator of the other. That is, if $\frac{a}{b} = \frac{c}{d}$, then $a \cdot d = b \cdot c$. This method creates a linear equation that can be solved using the algebraic methods discussed in previous sections.

Example 1: Solve the proportion $\frac{x}{5} = \frac{7}{10}$ for x.

Using cross-multiplication, we get $x \cdot 10 = 5 \cdot 7$. Simplifying the right-hand side, we have $10x = 35$. Dividing both sides by 10, we obtain $x = 3.5$.

```
x = 3.5
```

Applications of Proportions

Proportions are particularly useful in solving real-world problems involving rates, like speed or density, mixtures, and similarity in geometry.

Example 2: If a car travels 120 miles in 2 hours, what is its average speed?

Average speed can be computed using the ratio of distance to time. Thus, the speed s is given by the proportion $\frac{s}{1\,\text{hour}} = \frac{120\,\text{miles}}{2\,\text{hours}}$. Solving for s, we find that $s = 60$ mph.

```
Average speed = 60 mph
```

Example 3: A recipe for a cake requires 3 cups of flour for every 2 cups of sugar. If a baker wants to make a cake using 4 cups of sugar, how many cups of flour are needed?

Let f represent the cups of flour. The proportion here is $\frac{f}{4} = \frac{3}{2}$. Cross-multiplying yields $2f = 12$, so $f = 6$.

2.9. PROBLEMS INVOLVING PROPORTIONS AND RATIOS

```
Cups of flour needed = 6
```

Solving Problems with Ratios

When dealing with problems that involve ratios, it's often useful to express the ratios as fractions and then solve the resulting equations.

Example 4: In a classroom, the ratio of boys to girls is 3 to 4. If there are 28 students in total, how many boys and girls are there?

Let b represent the number of boys and g represent the number of girls. We are given that $\frac{b}{g} = \frac{3}{4}$ and $b + g = 28$. To solve this system of equations, express b in terms of g using the given ratio: $b = \frac{3}{4}g$. Substituting this into the second equation gives $\frac{3}{4}g + g = 28$. Solving for g yields $g = 16$, and consequently $b = 12$.

```
Number of boys = 12
Number of girls = 16
```

Understanding and applying the techniques for solving problems involving proportions and ratios equip students with tools to approach and solve a broad range of mathematical problems. Mastery of this topic fosters critical thinking skills that are beneficial not only in mathematics but in everyday reasoning and in various academic and professional fields.

2.10 Introduction to Linear Inequalities

Linear inequalities are mathematical statements that involve a comparison between two linear expressions. Unlike linear equations, which denote equality, inequalities express relationships such as greater than ($>$), less than ($<$), greater than or equal to ($\geq$), or less than or equal to ($\leq$). Linear inequalities in one variable take the general form $ax + b > 0$, $ax + b < 0$, $ax + b \geq 0$, or $ax + b \leq 0$, where a and b are real numbers, and x is the variable. Understanding how to solve and graph these inequalities is crucial for a variety of mathematical applications, including problem solving and modeling real-world scenarios.

One of the first concepts to grasp with linear inequalities is the method of solving them, which is similar to solving linear equations but with important distinctions related to the behavior of inequalities under certain operations. Specifically, when both sides of an inequality are multiplied or divided by a negative number, the direction of the inequality must be reversed to maintain the statement's truth. This property is a fundamental rule in the manipulation and simplification of inequalities.

To illustrate basic solution techniques, consider the inequality $2x - 5 > 3$. To solve for x, one would first add 5 to both sides of the inequality, resulting in $2x > 8$, and then divide both sides by 2, yielding $x > 4$. This solution means that any real number greater than 4 satisfies the original inequality.

Graphical representation of solutions is another critical aspect of understanding linear inequalities. When representing the solution set of an inequality on a number line,

2.10. INTRODUCTION TO LINEAR INEQUALITIES

a circle is used to indicate whether the boundary point is included in the solution set (closed circle for $\geq$ or $\leq$) or not (open circle for $>$ or $<$). For instance, the graphical representation of $x > 4$ includes an open circle at $x = 4$ and a shaded line extending to the right of 4, signifying all numbers greater than 4.

- The solution to a linear inequality in one variable is a range of values rather than a single value.

- Multiplying or dividing both sides of an inequality by a negative number reverses the direction of the inequality.

- Solutions to inequalities can be represented graphically on a number line, where open or closed circles are used to denote whether boundary points are included in the solution set.

Linear inequalities often appear in systems, similar to linear equations, where two or more inequalities are considered simultaneously. Solving systems of linear inequalities involves finding the set of values that satisfy all inequalities in the system. Graphically, this is represented as the intersection of the solution sets for each inequality, which can be particularly insightful for visualizing constraints in optimization problems and other applications.

Consider, for example, a system comprising two inequalities: $y \geq 2x + 1$ and $y < -x + 4$. The solution to this system involves graphing both inequalities on the same set of axes and identifying the region of the plane that satisfies both conditions simultaneously. This intersecting area, often referred to as the feasible region, represents the set of all possible solutions to the system.

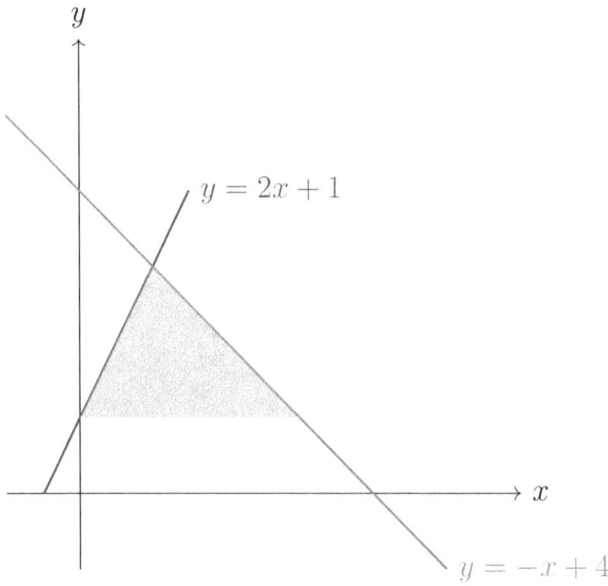

Linear inequalities serve as a powerful tool in algebra for describing and analyzing relationships between variables. Mastery of solving and graphically representing these inequalities lays the foundation for more advanced mathematical concepts and provides a robust framework for tackling real-world problems that involve constraints and conditions.

2.11 Solving Linear Inequalities

Solving linear inequalities is an essential skill that extends the concepts learned in solving linear equations. An inequality, unlike an equation, indicates that the two expressions involved are not necessarily equal, but instead, one is greater than, less than, greater than or equal to, or less than or equal to the other. The symbols used to denote these relationships are $>$, $<$, $\geq$, and $\leq$, respectively.

2.11. SOLVING LINEAR INEQUALITIES

The process of solving linear inequalities is similar to that of solving linear equations, with the primary goal being to isolate the variable on one side of the inequality. However, a notable difference arises when multiplying or dividing both sides of an inequality by a negative number, which results in the direction of the inequality symbol being reversed.

Consider a linear inequality in one variable, $ax + b < c$. The steps to solve it are as follows:

1. Begin by subtracting b from both sides of the inequality to get $ax < c - b$.

2. If a is a positive number, divide both sides by a to isolate x, yielding $x < \frac{c-b}{a}$.

3. If a is a negative number, divide both sides by a and reverse the inequality symbol, resulting in $x > \frac{c-b}{a}$.

Let's apply these steps to a specific example:

```
Solve the inequality: -3x + 5 > 2.

1. Subtract 5 from both sides: -3x > -3.
2. Divide by -3 and reverse the inequality: x < 1.
```

One of the key aspects of inequalities is expressing their solutions. Since inequalities often have multiple solutions, we typically represent these solutions on a number line or using interval notation. For the example above, the solution $x < 1$ can be represented on a number line with an open circle at 1 and a shaded line extending to the left to indicate all the numbers less than 1. In interval notation, this is written as $(-\infty, 1)$.

When solving inequalities involving absolute values, such as $|ax + b| < c$, the approach involves considering two cases due to the definition of absolute value. The first case equates the expression inside the absolute value to less than c, and the second case to greater than $-c$. This effectively removes the absolute value and creates two separate inequalities to solve.

```
Solve the inequality: |2x - 3| < 5.
1. Case 1: 2x - 3 < 5    -->    2x < 8     -->    x < 4.
2. Case 2: 2x - 3 > -5   -->    2x > -2    -->    x > -1.

Combining both cases gives the solution: -1 < x < 4.
```

Finally, when addressing inequalities with variables on both sides, the approach mirrors that of solving equations: simplify and collect like terms to isolate the variable on one side. Care must be taken when multiplying or dividing by negative quantities, remembering to reverse the inequality symbol when necessary.

Throughout this discussion, emphasis is placed on the correct representation of solutions, conceptual understanding of inequalities, and the procedural adjustments required when dealing with negative coefficients or absolute values. Mastery of these topics provides a solid foundation for further exploration into systems of inequalities and real-world applications, such as optimizing functions subject to constraints.

2.12 Graphical Representations of Solutions

Graphical representations provide a visual interpretation of the solutions to linear equations and inequalities, enhancing comprehension of the underlying mathematical

2.12. GRAPHICAL REPRESENTATIONS OF SOLUTIONS

concepts. This section explores the process of graphing linear equations and inequalities on a coordinate plane, interpreting these graphs, and using them to solve real-world problems.

To graph a linear equation, one must first understand that the solution to such an equation corresponds to a set of points on the Cartesian plane that form a straight line. This line represents all the pairs of (x, y) that satisfy the equation. The fundamental form of a linear equation is $y = mx + b$, where m is the slope of the line, indicating its steepness and direction, and b is the y-intercept, the point where the line crosses the y-axis.

For example, to graph $y = 2x + 1$:

- Start by plotting the y-intercept $(0, 1)$ on the graph.

- Use the slope, $m = 2$, to determine another point. From $(0, 1)$, move up 2 units (the rise) and 1 unit to the right (the run). This leads to the point $(1, 3)$.

- Plot the point $(1, 3)$ and draw a straight line through the two points.

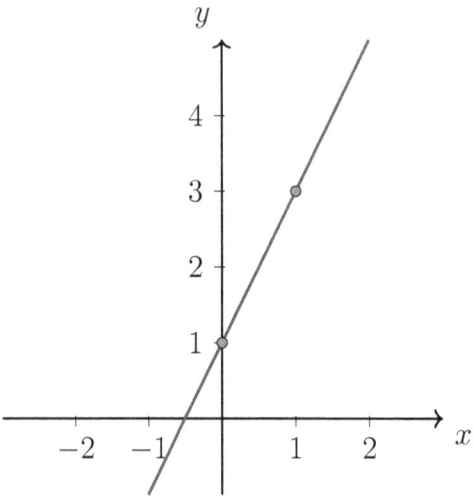

Graphing linear inequalities involves a similar process, with an additional consideration for the inequality sign. The solution set for a linear inequality is represented by a shaded area on one side of the boundary line, which corresponds to the graph of the associated linear equation. The boundary line is drawn as a solid line if the inequality includes equality (e.g., $\leq$ or $\geq$), indicating that points on the line are part of the solution set. It is drawn as a dashed line for strict inequalities (e.g., $<$ or $>$), indicating that points on the line are not part of the solution set.

Consider the inequality $y < 2x + 1$:

- Graph the line $y = 2x + 1$ as before, using a dashed line.

- Choose a test point not on the line, such as $(0, 0)$, and substitute it into the inequality. Since $0 < 2(0) + 1$, the area that contains $(0, 0)$ is shaded.

2.12. GRAPHICAL REPRESENTATIONS OF SOLUTIONS

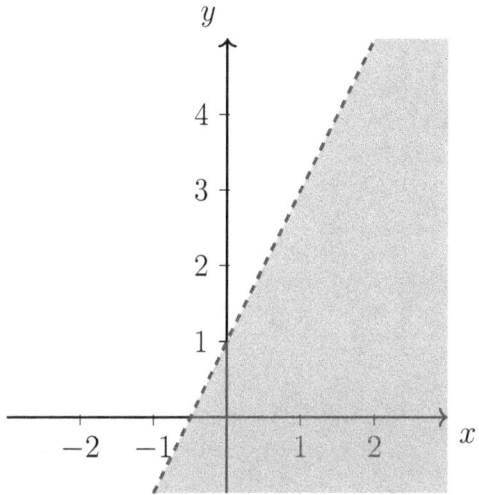

Graphical representations are not only effective for solving and understanding individual equations and inequalities but also offer valuable insights when dealing with systems of equations or inequalities. In the context of systems, the graphical approach enables the identification of intersecting points or overlapping regions that represent the solutions satisfying all equations or inequalities in the system.

In real-world applications, graphical methods provide a powerful tool for visualizing relationships and dependencies between variables, making abstract concepts tangible. For instance, in economics, graphs can illustrate the relationship between supply and demand, while in physics, they can represent the motion of objects.

Understanding the principles of graphical representations enriches the study of algebra, offering a multifaceted approach to problem-solving and analysis. Through practice and exploration, students can harness the power of graphs to interpret and solve mathematical challenges, laying a strong foundation for further mathematical studies and real-world decision-making.

Chapter 3

Solving Inequalities

This chapter focuses on understanding and solving various forms of inequalities, highlighting the differences and similarities between solving inequalities and solving equations. The discussion encompasses basic inequalities, multi-step inequalities, compound inequalities, and those involving absolute values. Additionally, it explores inequalities with quadratics and rational expressions, and introduces concepts like graphical solutions and linear programming. By emphasizing analytical and graphical methods, the chapter aims to equip students with the tools to address a broad range of problems involving inequalities, providing a solid foundation for more advanced mathematical studies.

3.1 Introduction to Inequalities

Inequalities are fundamental to mathematics, representing relationships between expressions that are not necessarily equal but rather less than, greater than, less than or equal to, or greater than or equal to each other. Unlike equations that denote equality, inequalities describe a range of possible values that satisfy the conditions set forth. This chapter introduces the concept of inequalities, providing the basis for understanding how they differ from equations and how they are solved and applied.

An inequality is written using one of the symbols < (less than), > (greater than), ≤ (less than or equal to), or ≥ (greater than or equal to). The expression on one side of the inequality is not strictly equal to the expression on the other side; instead, it falls within a range that the inequality defines.

$$a < b$$

The above inequality states that a is less than b. Here, a and b can be numbers, variables, or expressions. It is crucial to interpret this correctly: there are infinitely many numbers that could satisfy this inequality, rather than a single solution as is often sought in equations.

To visualize inequalities, one can plot them on a number line. For instance, the inequality $x < 3$ would be represented by an open circle at 3 on the number line, with a shaded line extending to the left, indicating all the numbers less than 3 are solutions to the inequality.

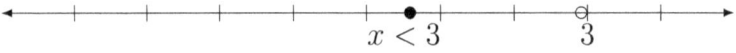

Inequalities can be manipulated similarly to equations, with operations such as adding, subtracting, multiplying, and dividing applied to both sides. However, there is a crucial distinction: multiplying or dividing both sides of an inequality by a negative number reverses the inequality symbol. This property is paramount to solving inequalities correctly.

$$-2x > 4$$
$$x < -2$$

This example illustrates that after dividing both sides by -2, the direction of the inequality is reversed, reflecting that x is less than -2 not greater than.

Solving inequalities is not just an academic exercise; it has practical applications in various fields such as engineering, economics, and science. Inequalities can model constraints and conditions in optimization problems, risk assessments, and decision-making processes. Understanding how to solve and graph inequalities is essential for interpreting and solving real-world problems.

The goal of this chapter is to develop a solid foundation in understanding and solving different types of inequalities. Starting with basic single-variable inequalities, the discussion will progress through multi-step, compound, and absolute value inequalities, setting the stage for exploring inequalities involving quadratics, rational expressions, and variables on both sides. By integrating analytical and graphical solutions, this chapter aims to equip students with a comprehensive set of tools for tackling a broad spectrum of inequality problems.

3.2 Solving Basic Inequalities

In this section, we focus on understanding and solving basic inequalities, a fundamental skill that underlies more complex topics in algebra. While the process shares some similarities with solving equations, there are notable differences, particularly in how inequality signs are treated.

An inequality, simply put, is a mathematical statement that one quantity is less than (or greater than) another. Basic inequalities involve at most one variable and do not require multiple steps to solve. They are typically of the form $ax + b < c$ or $ax + b > c$, where a, b, and c are constants.

To solve such inequalities, the goal is to isolate the variable on one side. This process often involves the following steps:

- Subtracting or adding the same number from/to both sides of the inequality.

- Dividing or multiplying both sides by the same positive number.

- Reversing the inequality sign when both sides are multiplied or divided by a negative number.

It is crucial to remember that unlike equations, multiplying or dividing by a negative number reverses the direction of the inequality sign. This rule is a key difference and must be carefully applied to ensure correct solutions.

Example 1: Consider solving the inequality $2x + 5 < 11$.

$$2x + 5 < 11$$
$$2x < 6 \quad \text{(Subtracting 5 from both sides)}$$
$$x < 3 \quad \text{(Dividing both sides by 2)}$$

The solution to this inequality is $x < 3$, which means that any value of x less than 3 makes the original inequality true.

3.2. SOLVING BASIC INEQUALITIES

Example 2: Consider solving the inequality $-3x + 4 \geq 1$.

$$-3x + 4 \geq 1$$
$$-3x \geq -3 \quad \text{(Subtracting 4 from both sides)}$$
$$x \leq 1 \quad \text{(Dividing both sides by -3 and reversing the inequality)}$$

The solution to this inequality is $x \leq 1$. It is crucial to notice the reversal of the inequality sign due to division by a negative number.

Graphical representation is an invaluable tool for understanding solutions to inequalities. To visualize the solution set of an inequality, one can graph a number line and shade the region representing all possible solutions. For instance, the solution to $x < 3$ can be represented with a shaded area to the left of 3 on the number line and an open circle at 3, indicating that 3 is not included in the solution set.

This graphical solution provides a visual understanding of the range of values that satisfy the original inequality.

Solving basic inequalities is a key skill in algebra. By carefully applying the rules for handling inequality signs, especially when multiplying or dividing by negative numbers, students can accurately find solutions to these problems. Furthermore, graphical representations offer an intuitive way of understanding and visualizing the solutions, reinforcing the analytical techniques.

3.3 Solving Multi-Step Inequalities

Multi-step inequalities, akin to multi-step equations, necessitate a series of operations to isolate the variable and find its range of values. This section delineates the methodology for solving such inequalities, illustrating the process with examples and emphasizing the preservation of inequality's directionality.

When addressing multi-step inequalities, the primary objective is to simplify the inequality step-by-step until the variable is by itself on one side. Unlike equations, which yield a specific value or set of values as solutions, inequalities render a range or sets of values that satisfy the condition.

Key Principles:

- *Operation Consistency:* Similar to solving equations, whatever operation (addition, subtraction, multiplication, or division) you perform on one side of an inequality, you must also perform on the other side.

- *Inequality Reversal:* A critical distinction when solving inequalities is that if you multiply or divide both sides by a negative number, you must reverse the direction of the inequality.

Solving Process:

The general steps for solving multi-step inequalities are as follows:

1. Simplify both sides of the inequality, if possible, by combining like terms and applying the distributive property.

2. Get all variable terms on one side and constants on the other side of the inequality.

3. Isolate the variable by performing appropriate operations on both sides of the inequality.

4. If the operation involves multiplying or dividing by a negative number, reverse the direction of the inequality.

5. Verify the solution by testing a value from the solution set in the original inequality.

Examples:

Consider the inequality $3(2x - 4) > 18$.

1. We start by applying the distributive property: $6x - 12 > 18$.

2. Next, isolate variable terms by adding 12 to both sides: $6x > 30$.

3. Finally, divide both sides by 6, obtaining $x > 5$.

4. Verification can be done by testing a value greater than 5, say 6, in the original inequality: $3(2 \cdot 6 - 4) = 3(12 - 4) = 3 \cdot 8 = 24 > 18$, which is true.

```
Solution: x > 5
```

Let us consider another example involving a negative coefficient: $-2(x + 3) \leq 4 - 2x$.

1. Apply the distributive property: $-2x - 6 \leq 4 - 2x$.

2. Observing that $-2x$ terms cancel out when moved to one side, we simplify: $-6 \leq 4$.

3. Since there is no variable left to solve for, we interpret the result directly. The statement $-6 \leq 4$ is always true; hence, the solution is all real numbers.

In dealing with multi-step inequalities, several problems might arise, varying from the cancellation of variable terms leading to a trivial true or false statement, to managing inequalities with absolute values or rational expressions, which are addressed in subsequent sections.

Solving multi-step inequalities requires a methodical approach, beginning with simplification and manipulation of the inequality until the variable is isolated. Critical to this process is the adjustment of the inequality direction when multiplying or dividing by negative numbers. Through practice, applying these principles becomes intuitive, allowing for efficient solving of complex, real-world inequality problems.

3.4 Compound Inequalities

Compound inequalities are expressions involving two separate inequalities that are connected by the words "and" or "or". Understanding and solving compound inequalities is a vital skill in algebra since they frequently occur in various mathematical contexts and real-world situations. This section focuses on the two types of compound inequalities, how to solve them, and how to graph their solutions.

3.4. COMPOUND INEQUALITIES

Types of Compound Inequalities

There are two main types of compound inequalities:

- "And" inequalities (Conjunctions): These are composed of two inequalities that must both be true at the same time. An example is $x > 1$ and $x < 5$. For a value to be a solution to the compound inequality, it must satisfy both conditions simultaneously.

- "Or" inequalities (Disjunctions): These are composed of two inequalities where at least one of the inequalities must be true. An example is $x < -2$ or $x > 3$. A value is considered a solution to the compound inequality if it satisfies at least one of the conditions.

Solving "And" Compound Inequalities

To solve "and" compound inequalities, one must find the set of all values that satisfy both inequalities simultaneously. The solution set is often an interval on the number line.

Example:

Solve the compound inequality $1 < x - 3 \leq 5$.

$$1 < x - 3$$
$$1 + 3 < x - 3 + 3$$
$$4 < x$$

And

$$x - 3 \leq 5$$
$$x - 3 + 3 \leq 5 + 3$$
$$x \leq 8$$

Combining the two parts, we get $4 < x \leq 8$. The solution set consists of all real numbers greater than 4 and less than or equal to 8.

Solving "Or" Compound Inequalities

To solve "or" compound inequalities, find the union of the solution sets of each inequality. The solution to the compound inequality is any value that makes at least one of the inequalities true.

Example:

Solve the compound inequality $x + 2 \leq -3$ or $x - 4 > 1$.

$$x + 2 \leq -3$$
$$x \leq -5$$

And

$$x - 4 > 1$$
$$x > 5$$

The solution set is the union of $x \leq -5$ and $x > 5$. It includes all real numbers less than or equal to -5 or greater than 5.

3.4. COMPOUND INEQUALITIES

Graphical Representation of Compound Inequalities

A graphical representation on a number line can effectively display the solution set of compound inequalities.

- For "and" inequalities, shade the interval between the points that satisfy both conditions, including or excluding endpoints based on the inequality signs.

- For "or" inequalities, shade the intervals or rays that satisfy at least one of the conditions, with the same consideration for endpoints as above.

```
Example of Graphing "And" Compound Inequality (4 < x <= 8):

------------------------------o========o--------------------------
                              4        8
Explanation: The open circle at 4 indicates that 4 is not included in
the solution set, while the closed circle at 8 indicates that 8 is
included.


Example of Graphing "Or" Compound Inequality (x <= -5 or x > 5):

-----------•=================]-------[=================•-----------
           -5                         5
Explanation: The closed circle at -5 and to the left indicates that -5
and all numbers less than -5 are included. The open circle at 5 and
to the right indicates that all numbers greater than 5 are included.
```

Understanding and solving compound inequalities are crucial for tackling a variety of algebraic problems. This section has introduced the core concepts of compound inequalities, providing a basis for further exploration in algebra and its applications in solving real-world problems.

3.5 Absolute Value Inequalities

Solving absolute value inequalities is an essential skill in algebra that combines the understanding of absolute value and inequality properties. The absolute value of a number measures its distance from zero on the number line, irrespective of the direction. Consequently, absolute value inequalities may have two distinct cases to consider depending on whether the inequality is "less than" or "greater than".

Definition of Absolute Value Inequalities

An absolute value inequality is an inequality that contains an absolute value expression. The basic form of such inequalities can be expressed as $|A| < B$, $|A| > B$, $|A| \leq B$, or $|A| \geq B$, where A is an algebraic expression, and B is a non-negative real number.

Solving Absolute Value Inequalities of the Form $|A| < B$ and $|A| \leq B$

To solve inequalities of the form $|A| < B$ or $|A| \leq B$, where $B > 0$, one must recognize that the solution involves finding all values of the variable within a distance B from zero.

$$|A| < B \implies -B < A < B$$
$$|A| \leq B \implies -B \leq A \leq B$$

These inequalities state that A lies between $-B$ and B.

3.5. ABSOLUTE VALUE INEQUALITIES

Example 1: Solve $|x - 3| < 4$.

$$-4 < x - 3 < 4$$
$$-4 + 3 < x < 4 + 3$$
$$-1 < x < 7$$

Solution: x \in (-1, 7)

Solving Absolute Value Inequalities of the Form $|A| > B$ and $|A| \geq B$

Inequalities of the form $|A| > B$ or $|A| \geq B$ involve finding all values of the variable that are more than a distance B from zero. This results in two separate intervals, as the solution lies either to the left of $-B$ or to the right of B.

$$|A| > B \implies A > B \text{ or } A < -B$$
$$|A| \geq B \implies A \geq B \text{ or } A \leq -B$$

Example 2: Solve $|x + 2| > 3$.

$$x + 2 > 3 \text{ or } x + 2 < -3$$
$$x > 1 \text{ or } x < -5$$

Solution: x \in (-\infty, -5) \cup (1, \infty)

Graphical Interpretation

The solutions to absolute value inequalities can also be represented graphically on the number line. This is particularly useful for visual learners and adds an additional layer of understanding to the algebraic solution.

The arrows indicate the direction of the inequality solution on the number line, while the filled and unfilled circle indicates whether the endpoint is included in the solution set.

Understanding how to manipulate and solve absolute value inequalities is a powerful tool in algebra. These inequalities often appear in various mathematical and real-world problems. The strategy for solving them depends critically on understanding whether the inequality is of a "less than" or "greater than" form, which dictates the nature of its solution set. By grasitating these concepts, students will significantly enhance their ability to tackle problems that involve distances and ranges both on the real number line and in real-life situations.

3.6 Inequalities Involving Quadratics

Inequalities involving quadratics are a critical area of study as they extend our understanding beyond linear relationships, enabling us to solve a broader range of real-world problems. This section focuses on techniques for solving quadratic inequalities, which typically take the form $ax^2 + bx + c > 0$, $ax^2 + bx + c < 0$, $ax^2 + bx + c \geq 0$, or $ax^2 + bx + c \leq 0$, where a, b, and c are constants with $a \neq 0$.

The process of solving quadratic inequalities closely mir-

3.6. INEQUALITIES INVOLVING QUADRATICS

rors that of solving quadratic equations; however, the solution to a quadratic inequality is often an interval or a set of intervals rather than discrete values.

Step 1: Solve the Corresponding Quadratic Equation

Begin by solving the corresponding quadratic equation $ax^2 + bx + c = 0$. This can be done by factoring, completing the square, or applying the quadratic formula. The solutions to this equation, called the *critical values*, will partition the real number line into intervals.

Example: Consider the inequality $x^2 - 3x - 4 < 0$.

First, solve the equation $x^2 - 3x - 4 = 0$.

x = 4, x = -1

These critical values divide the number line into three intervals: $(-\infty, -1)$, $(-1, 4)$, and $(4, \infty)$.

Step 2: Test the Intervals

Select a test point from each interval and substitute it back into the original inequality to determine whether the interval satisfies the inequality.

Example Continued:

Testing the interval $(-\infty, -1)$ with $x = -2$:

$$x^2 - 3x - 4 < 0$$
$$(-2)^2 - 3(-2) - 4 < 0$$
$$4 + 6 - 4 < 0$$
$$6 < 0, \text{ which is false.}$$

Testing the interval $(-1, 4)$ with $x = 0$:

$$x^2 - 3x - 4 < 0$$
$$(0)^2 - 3(0) - 4 < 0$$
$$-4 < 0, \text{ which is true.}$$

Testing the interval $(4, \infty)$ with $x = 5$:

$$x^2 - 3x - 4 < 0$$
$$(5)^2 - 3(5) - 4 < 0$$
$$25 - 15 - 4 < 0$$
$$6 < 0, \text{ which is false.}$$

Step 3: Combine the Results

Based on the test results, the solution to the original inequality $x^2 - 3x - 4 < 0$ is the interval $(-1, 4)$.

Graphical Interpretation

Graphically, solving a quadratic inequality involves identifying the regions where the graph of the quadratic expression lies above or below the x-axis. The roots of the corresponding quadratic equation indicate where the graph intersects the x-axis, and testing intervals determines the sections of the graph that satisfy the inequality.

3.6. INEQUALITIES INVOLVING QUADRATICS

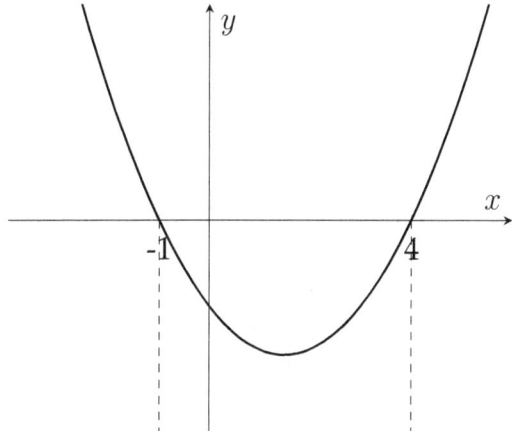

The dashed lines at $x = -1$ and $x = 4$ represent the critical values. The segment of the graph between these two points lies below the x-axis, indicating that the values of x in this interval satisfy the inequality $x^2 - 3x - 4 < 0$.

Practice Problems

Given your understanding of solving quadratic inequalities, try the following problems:

- Solve the inequality $x^2 + x - 6 \geq 0$.

- Find the solution set for the inequality $2x^2 - 8x + 6 < 0$.

- Determine the interval(s) in which the quadratic expression $3x^2 + 12x + 7 > 0$ holds true.

In summary, solving quadratic inequalities requires identifying the critical values by solving the corresponding quadratic equation, testing the intervals created by these critical values, and determining which intervals satisfy the original inequality. This technique is essential for analyzing situations modeled by quadratic relationships and facilitates a deeper understanding of the behavior of quadratic functions.

3.7 Rational Inequalities

Rational inequalities involve ratios of polynomial expressions. Unlike equations, where equality is asserted, inequalities indicate a relation involving less than (<), greater than (>), less than or equal to (≤), or greater than or equal to (≥). The process of solving rational inequalities is systematic and requires careful analysis of the expressions involved. This section will guide you through the method to solve these inequalities, emphasizing the critical steps necessary for a comprehensive understanding.

When dealing with rational inequalities, the primary goal is to find the x-values that make the inequality true. The method involves several key steps, starting with simplifying the inequality if necessary, finding the critical values, testing intervals defined by these critical values, and finally, interpreting these tests to solve the inequality.

Consider a general rational inequality of the form:

$$\frac{P(x)}{Q(x)} \succ 0$$

where $P(x)$ and $Q(x)$ are polynomial functions, and $\succ$ denotes any of the four inequality symbols.

The following detailed steps will guide you through solving rational inequalities effectively:

Step 1: Simplify the Inequality If possible, simplify the inequality by canceling common factors in the numerator

3.7. RATIONAL INEQUALITIES

and denominator. However, be cautious of factors that may introduce undefined values or affect the inequality's direction.

Step 2: Determine the Critical Values Critical values are x-values that make the numerator zero or the denominator zero. These values are obtained by setting $P(x) = 0$ and $Q(x) = 0$, which leads to the critical points when the inequality changes its truth value. Solving these equations for x gives the critical values:

$$P(x) = 0 \Rightarrow x = x_1, x_2, \ldots, x_n$$
$$Q(x) = 0 \Rightarrow x = y_1, y_2, \ldots, y_m$$

Step 3: Plot the Critical Values on a Number Line Mark the critical values obtained from Step 2 on a number line. These values partition the number line into intervals. It's crucial to consider open or closed circles based on whether the inequality is strict (<, >) or non-strict ($\leq$, $\geq$).

Step 4: Test Points in the Intervals Select test points from each of the intervals created by the critical values. Substitute these test points into the original inequality to determine if the inequality holds. Record whether each interval satisfies the inequality.

Step 5: Write the Solution The solution to the inequality consists of all intervals that satisfy the inequality. Express the solution in interval notation, paying attention to whether the endpoints (critical values) are included or excluded based on the nature of the inequality.

Example: Solve the rational inequality:

$$\frac{x-2}{x^2 - x - 6} > 0$$

Solution: First, factor the denominator as $x^2 - x - 6 = (x-3)(x+2)$. The critical values are found by setting the numerator and the factored denominator equal to zero:

$$x - 2 = 0 \Rightarrow x = 2$$
$$x - 3 = 0 \Rightarrow x = 3$$
$$x + 2 = 0 \Rightarrow x = -2$$

Plot these values on a number line and select test points in the intervals $(-\infty, -2)$, $(-2, 2)$, $(2, 3)$, and $(3, \infty)$. Test these points in the original inequality to determine where the inequality is satisfied.

After testing, we find that the inequality is satisfied in the intervals $(-\infty, -2)$ and $(2, 3)$. Therefore, the solution in interval notation is:

(-\infty, -2) U (2, 3)

Understanding and solving rational inequalities requires careful attention to the procedure detailed above. By identifying critical values and testing intervals, you can systematically determine the x-values that satisfy a given rational inequality. This method not only aids in solving rational inequalities effectively but also strengthens your overall mathematical problem-solving skills.

3.8 Inequalities with Variables on Both Sides

Inequalities with variables on both sides present a slightly more complex challenge than the inequalities we have encountered thus far. Nevertheless, the methods used to solve such inequalities mirror those employed in handling equations, with careful consideration of the direction of the inequality sign.

The general form of these inequalities can be expressed as:
$$ax + b < cx + d$$
or in the context of non-strict inequalities,
$$ax + b \leq cx + d$$
where a, b, c, and d are constants, and x is the variable. The goal is to isolate x on one side of the inequality to determine its values satisfying the inequality.

Step 1: Simplify both sides if necessary. This involves expanding expressions and combining like terms which sets a clearer path to isolating the variable.

Step 2: Move all terms involving x to one side of the inequality. This might require adding or subtracting terms from both sides. It is crucial to remember that, unlike in equations, multiplying or dividing both sides of an inequality by a negative number reverses the direction of the inequality sign.

Step 3: Isolate the variable. Once variables are on one side and constants on the other, divide or multiply to solve for x.

Example:

Consider the inequality
$$3x + 4 < 5x - 2$$

Step 1: This inequality is already simplified.

Step 2: Subtract $3x$ from both sides to move variable terms to one side:
$$4 < 2x - 2$$

Then, add 2 to both sides of the inequality:
$$6 < 2x$$

Step 3: Divide both sides by 2 to solve for x:
$$3 < x$$

Thus, x must be greater than 3 to satisfy the original inequality.

Consideration when Multiplying or Dividing by Negative Numbers:

If, in the process of isolating x, one must multiply or divide both sides by a negative number, it's crucial to reverse the inequality sign. This operation can seem counterintuitive, but it is vital to ensuring the correct solution.

Example:
$$-2x + 3 > 1 - x$$

First, let's collect x-terms on the left side and constant terms on the right by adding $2x$ and subtracting 1 on both sides:
$$-2x + 2x + 3 - 1 > 1 - 1 - x + 2x$$
$$2 > x$$

In summary, the inequality $2 > x$, or equivalently $x < 2$, indicates that the solution set consists of all numbers less than 2.

Checking Solutions:

It's wise to verify solutions to these inequalities by testing values in the original inequality. Since inequalities can have multiple valid solutions, check values on both sides of the critical point(s) identified during the solving process.

3.9. GRAPHICAL SOLUTIONS OF INEQUALITIES

```
Example Check for x < 2:
Test x = 0: -2(0) + 3 > 1 - 0 -> 3 > 1 (True)
Test x = 3: -2(3) + 3 > 1 - 3 -> -3 > -2 (False)
```

Through this process, we hold that our solution $x < 2$ is indeed correct, supported by our tests where values of x on one side of 2 maintain the inequality's truth, while those on the other side do not.

Particularly, in the sphere of inequalities with variables on both sides, astute manipulation of terms, attention to directionality when engaging with negatives, and verification through substitution are essential to mastering these problems. As we proceed, remember that practice with varied examples will embed these critical steps, enabling adept handling of increasingly complex inequalities.

3.9 Graphical Solutions of Inequalities

Inequalities can often be understood and solved more intuitively by representing them graphically on the coordinate plane. This approach not only offers a visual interpretation of the solution set but also facilitates the comparison and analysis of multiple inequalities simultaneously. This section will elucidate the process of graphing linear and non-linear inequalities, followed by the method to find the solution set of system of inequalities.

Graphing Linear Inequalities

Linear inequalities are inequalities that can be written in one of the following forms: $ax + by < c$, $ax + by > c$, $ax + by \leq c$, or $ax + by \geq c$, where a, b, and c are constants. The graphical solution of a linear inequality represents all the points (x, y) that satisfy the inequality.

To graph a linear inequality, follow these steps:

1. Firstly, graph the boundary line, which is obtained by treating the inequality as an equality, i.e., by replacing the inequality sign with an equality sign ($=$). If the original inequality involves $\leq$ or $\geq$, draw a solid line to represent the boundary; if it involves $<$ or $>$, draw a dashed line to indicate that points on the line are not included in the solution set.

2. Next, choose a test point that is not on the boundary line, such as the origin $(0, 0)$, unless the boundary line passes through the origin. Substitute the coordinates of the test point into the inequality.

3. If the test point satisfies the inequality, shade the region of the plane containing the test point. Otherwise, shade the opposite side.

```
Example: Graph the inequality y < 2x + 1.
```

Graphing Non-Linear Inequalities

Non-linear inequalities involve quadratic functions, absolute value functions, or other non-linear functions. The process for graphing these inequalities is similar to linear inequalities but with an emphasis on identifying the curve that serves as the boundary.

3.9. GRAPHICAL SOLUTIONS OF INEQUALITIES

For quadratic inequalities of the form $ax^2 + bx + c < 0$ or $ax^2 + bx + c > 0$, the boundary is a parabola. The steps for graphing these inequalities are as follows:

1. Graph the parabola by converting the inequality to an equation.

2. Determine whether the parabola opens upwards or downwards by observing the sign of a. If $a > 0$, it opens upwards; if $a < 0$, it opens downwards.

3. Use a test point to determine which side of the parabola to shade. Remember that for $<$ or $>$ inequalities, the parabola is dashed, indicating that points on it are not part of the solution.

Example: Graph the inequality x^2 - 4x - 5 < 0.

Systems of Inequalities

A system of inequalities consists of two or more inequalities with the same variables. The solution set of the system is the intersection of the solution sets of the individual inequalities, i.e., the region that satisfies all inequalities simultaneously.

To graph a system of inequalities:

1. Graph each inequality on the same coordinate plane following the steps outlined previously. Remember to identify the boundary and shading direction for each inequality accurately.

2. The solution set is where the shaded regions overlap. This is the region that satisfies all inequalities in the system.

```
Example: Find the solution set of the system consisting of y < 2x + 1
and y > x - 2.
```

By visually representing inequalities, students can more easily grasp the concept of solution sets and how they vary with the type of inequality. Graphical solutions also facilitate understanding the impact of including or excluding boundary points and the intersection of solution sets in systems of inequalities. Mastery of these graphical methods will serve as a strong foundation for solving more complex algebraic problems and applications in optimization and linear programming.

3.10 Systems of Inequalities

A system of inequalities consists of two or more inequalities that are solved simultaneously. Solving such systems is crucial in identifying common solutions that satisfy all inequalities within the system. This section delves into methods of solving systems of linear inequalities, drawing parallels to solving systems of linear equations, but with notable distinctions in approach and representation of solutions.

A linear inequality in two variables, such as x and y, can be expressed in the form $ax+by < c$, $ax+by \leq c$, $ax+by > c$, or $ax + by \geq c$, where a, b, and c are constants. The solutions to these inequalities are represented as regions on the Cartesian plane, rather than as points.

3.10. SYSTEMS OF INEQUALITIES

Graphical Solution of Systems of Inequalities

The graphical method is a powerful tool for visualizing and solving systems of inequalities. The steps involved are as follows:

1. Begin by graphing each inequality on the same coordinate plane. To graph an inequality, first, graph the boundary line by converting the inequality to an equation (replacing $<$, $\leq$, $>$, or $\geq$ with $=$). If the inequality is strict ($<$ or $>$), the boundary line is dashed, indicating that points on the line are not part of the solution set. If the inequality includes the equality ($\leq$ or $\geq$), the line is solid.

2. Determine the solution region for each inequality. This is done by selecting a test point that is not on the boundary line (often, the origin $(0,0)$ is a convenient choice) and substituting this point into the inequality. If the inequality is true, the half-plane containing the test point represents the solution set. If false, the opposite half-plane is the solution set.

3. Shade the solution region for each inequality. Where the shaded regions overlap, indicates the common solutions to the system.

4. The intersection of all shaded regions is the solution set of the system of inequalities. This solution set can be represented graphically as a region (bounded or unbounded) on the coordinate plane.

It is essential to visualize these steps through examples.

Example: Consider the system of inequalities:

$$y \geq x - 2$$
$$y < 2x + 1$$

First, graph the lines $y = x - 2$ and $y = 2x + 1$ where the former has a solid line and the latter a dashed line. To determine the solution region of $y \geq x - 2$, test the point $(0, 0)$. Since $0 \geq -2$, the region containing the origin is shaded. For $y < 2x + 1$, using the same test point, since 0 is not less than 1, the region opposite the origin is shaded for this inequality. The solution set is where these regions overlap.

Graphical Representation:

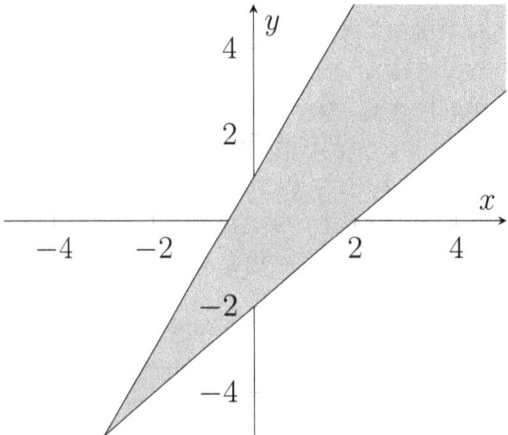

The shaded region above is the graphical representation of the solution set to the system of inequalities.

Applications of Systems of Inequalities

Systems of inequalities are widely used in various fields, including economics, engineering, and operations research, particularly in optimization and feasibility studies. For instance, in linear programming, a type of optimization problem, the objective is to maximize or minimize a linear function subject to a system of linear inequalities, representing constraints on the variables. This highlights the practical significance of understanding how to solve and graph systems of inequalities.

The method of graphing systems of inequalities involves plotting each inequality separately and then determining the overlapping shaded regions that satisfy all inequalities simultaneously. This approach not only aids in visualizing the solution set but also serves as a foundation for more advanced topics in mathematics and its applications in real-world problems.

3.11 Applications of Inequalities in Word Problems

The application of inequalities in word problems is a pivotal skill that stretches beyond the confines of a mathematics classroom, finding utility in various real-world scenarios such as economics, engineering, and everyday decision making. This section elucidates the process of translating textual information into mathematical inequalities and demonstrates systematic approaches to solving these problems.

The first step in addressing inequalities in word problems is the identification and definition of variables. Variables represent quantities that vary, and their proper selection is foundational to constructing meaningful expressions that mirror real-life situations.

- Identify what the question is asking. This will typically end up being your variable.

- Determine what inequalities represent the constraints or conditions mentioned in the problem.

- Translate the word problem into one or more inequalities.

Once the variables are defined, and the inequalities are set up, the next step involves solving the inequalities using methods discussed in preceding sections, which include simplification, handling multi-step inequalities, applying the principle of addition or subtraction, and dealing with multiplication or division, especially noting the rules regarding the inversion of the inequality sign when multiplying or dividing by a negative number.

Example 1: A school club is organizing a fundraiser. They need to raise at least $500 but cannot spend more than $200 in expenses. If x represents the amount of money raised and y represents the expenses, express this situation with inequalities.

Solving this involves setting up two inequalities based on the conditions provided:

$$x \geq 500$$
$$y \leq 200$$

Example 2: A company decides to produce a minimum of 100 units and a maximum of 300 units of a new product. The production cost per unit is $15, and the budget allocated for production is not to exceed $4000. If x represents the number of units produced and C represents the total production cost, formulate the inequalities representing these conditions.

3.11. APPLICATIONS OF INEQUALITIES IN WORD PROBLEMS

First, the production constraint translates to:

$$100 \leq x \leq 300$$

The budget constraint translates to:

$$C = 15x$$

Given that $C \leq 4000$, substituting for C yields:

$$15x \leq 4000$$

After setting up the inequalities, solving them involves standard algebraic manipulations. For both examples, the solution set describes the feasible region that meets the problem's conditions. Graphical methods can also be utilized to represent these constraints visually, providing an intuitive understanding of the feasible region where all conditions are satisfied simultaneously.

Real-world problems often involve multiple constraints leading to systems of inequalities, which can be solved graphically or algebraically to find the solution set that satisfies all the conditions. Whether determining the number of items a business can produce given budgetary constraints, optimizing resource allocation, or deciding on the best course of action given certain limitations, inequalities are invaluable tools in translating complex real-world problems into manageable mathematical terms.

To reinforce understanding, students should engage with a variety of problems, practicing the identification of variables and the translation of textual information into mathematical inequalities. This process not only enhances algebraic skills but also hones critical thinking and problem-solving abilities, underscoring the relevance of mathematics in analyzing and making decisions about real-world situations.

3.12 Introduction to Linear Programming

Linear programming is a mathematical technique used to find the best possible solution to an optimization problem within a given set of constraints. It involves maximizing or minimizing a linear objective function, subject to a system of linear inequalities that represent the constraints on the problem.

The objective function is a linear equation that represents the decision criteria of the problem. It could be the profit to maximize or the cost to minimize, depending on the context of the problem. The constraints, also linear inequalities, define the feasible region within which the solution must lie. This feasible region is a convex polygon or polyhedron in two or more dimensions.

The foundation of solving any linear programming problem is understanding the graphical method for two-variable problems. This method provides a visual interpretation of the feasible region and the objective function.

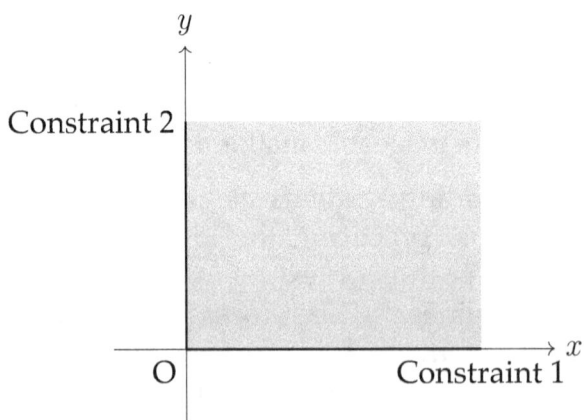

The figure above represents a simple linear programming

problem with two constraints and a feasible region. Every point in this region is a possible solution to the problem, but the optimal solution—where the objective function achieves its maximum or minimum value—will lie on the boundary of this region.

To identify this optimal point, the graphical method involves:

- Plotting the constraints on a graph to define the feasible region.
- Graphically representing the objective function as a line that can move parallel to itself.
- Moving the objective function line within the feasible region to find the point at which it touches the boundary last when maximizing (or first when minimizing), indicating the optimal solution.

The Corner Point Theorem is an important concept in linear programming, stating that if an optimal solution exists, it will be found at a vertex (corner point) of the feasible region. Thus, by evaluating the objective function at each vertex, we can determine the optimal solution without inspecting every point within the region.

For problems involving more than two variables, the graphical method is not feasible, and we turn to algebraic methods, such as the Simplex algorithm, designed to handle multidimensional linear programming problems efficiently.

Linear programming applications are widespread, ranging from business and economics, where it is used for profit maximization and cost minimization, to engineering for optimizing design processes, and even in health care for resource allocation.

An example problem can elucidate the application of linear programming:

```
Maximize P = 3x + 4y
Subject to:
1.    x + 2y   8
2.    3x + y   9
3.    x   0, y   0
```

The constraints define the feasible region, and by applying the graphical method or an algebraic approach, we seek to find the values of x and y that maximize P under these constraints.

Linear programming, though a specific field, illustrates the power of mathematical modeling in decision-making processes. With advancements in computational algorithms, the feasibility of solving large-scale linear programming problems has significantly improved, further broadening the scope of its applications.

In summary, linear programming offers a powerful toolkit for solving optimization problems in various settings. Understanding its conceptual framework and mastering its solution methods paves the way for tackling more complex problems with confidence.

Chapter 4

Graphing Linear Equations and Inequalities

This chapter explores the graphical representation of linear equations and inequalities, starting with the basics of the coordinate plane and progressing to more advanced topics such as slope, various forms of linear equations, and how to graph linear inequalities. The emphasis is on understanding the relationships between algebraic expressions and their graphical counterparts. Through detailed explanations and examples, students will learn how to visually interpret and solve problems involving linear equations and inequalities, enhancing their overall comprehension of algebraic concepts and their practical applications.

4.1 The Coordinate Plane

The coordinate plane, also known as the Cartesian plane, forms the foundational basis for graphing linear equations and inequalities. It is a two-dimensional surface defined by two perpendicular lines: the x-axis (horizontal) and the y-axis (vertical). These axes intersect at a point known as the origin, denoted by the coordinates (0, 0). The purpose of the coordinate plane is to provide a visual representation of numerical relationships, enabling the plotting of points, lines, and curves corresponding to mathematical expressions.

The coordinate plane is divided into four quadrants by its axes. These are numbered counter-clockwise starting from the quadrant in the upper right. Quadrant I contains points where both x and y coordinates are positive, Quadrant II contains points where x is negative and y is positive, Quadrant III contains points where both x and y are negative, and Quadrant IV, where x is positive and y is negative.

Points on the plane are represented by ordered pairs (x, y). The first number, x, indicates the position along the horizontal axis, while the second number, y, reveals the position along the vertical axis. To plot a point, one starts at the origin, moves horizontally to the x-coordinate, and then vertically to the y-coordinate.

```
Example: Plotting the Point (3, -2)
1. Start at the origin (0, 0).
2. Move 3 units to the right (along the x-axis).
3. Move 2 units down (along the y-axis).
The resulting point is in Quadrant IV.
```

Each point on the coordinate plane corresponds uniquely to an ordered pair, and each ordered pair identifies a

4.1. THE COORDINATE PLANE

unique point. This one-to-one correspondence enables the precise graphing and analysis of mathematical relationships.

For example, the point P(4, 5) is 4 units to the right of the origin and 5 units above it.

The distance between points on the coordinate plane can be calculated using the Pythagorean theorem, and the midpoint of a segment connecting two points can be found by averaging their x-coordinates and y-coordinates.

$$\text{Distance formula:} \quad d = \sqrt{(x_2 - x_1)^2 + (y_2 - y_1)^2}$$

$$\text{Midpoint formula:} \quad M = \left(\frac{x_1 + x_2}{2}, \frac{y_1 + y_2}{2}\right)$$

These concepts form the basis for understanding more complex geometric and algebraic topics, such as the slope of a line and the equations representing lines and curves.

Lines on the coordinate plane can be graphed using various forms of linear equations. The simplest is the graph of a horizontal line, $y = k$, where k is a constant and the line passes through all points with y-coordinate k. Conversely, a vertical line is represented by $x = h$, where h is a constant and the line passes through all points with x-coordinate h.

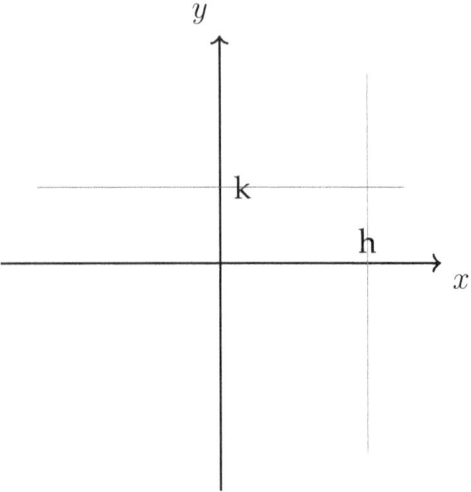

The coordinate plane is a vital tool in mathematics for visually representing and analyzing relationships between numbers. By understanding how to navigate this plane, plot points, and interpret the positions of lines and shapes, students gain a deeper insight into the graphical aspects of algebra.

4.2 Graphing Linear Equations in Two Variables

Graphing linear equations in two variables is a fundamental aspect of understanding algebra and its geometric interpretations. This section will explore the process of graphing linear equations by elucidating the roles of the coordinate plane, the concept of a linear equation, and the step-by-step procedure for graphing.

The foundation of graphing a linear equation in two variables lies in the understanding of the coordinate plane. The coordinate plane, also referred to as the Cartesian

4.2. GRAPHING LINEAR EQUATIONS IN TWO VARIABLES

plane, consists of two perpendicular number lines that intersect at their zero points. These number lines are known as axes, where the horizontal axis is called the x-axis and the vertical axis is the y-axis. The point of intersection of these axes is known as the origin, marked as (0, 0). Each point in the coordinate plane represents an ordered pair (x, y), showcasing the position of that point relative to the origin.

A linear equation in two variables typically takes the form $ax + by = c$, where a, b, and c are constants, and x and y are variables. The graph of such an equation is always a straight line. Therefore, the primary objective in graphing a linear equation is to draw that line on the coordinate plane. To accomplish this, we focus on finding two or more points on the line, determining the position of these points, and then connecting them with a straight line.

Example: Consider the equation $2x + y = 4$. To graph this equation, follow these steps:

1. **Choose values for x:** Start by selecting convenient values for x to find corresponding values of y. It is often helpful to choose values such as -1, 0, and 1 because they simplify calculations.

2. **Solve for y:** Substitute the chosen values of x into the equation to solve for y.

    $When\ x = -1,\ 2(-1) + y = 4,\ hence\ y = 6.$
    $When\ x = 0,\ 2(0) + y = 4,\ hence\ y = 4.$
    $When\ x = 1,\ 2(1) + y = 4,\ hence\ y = 2.$

3. **Plot the points:** Plot the points obtained from the above step on the coordinate plane.

    $(-1, 6),\ (0, 4),\ and\ (1, 2)$

4. **Draw the line:** Connect the plotted points with a straight line. This line is the graph of the equation $2x + y = 4$.

To illustrate the resulting graph, one may use a set of axes marked with the plotted points and the drawn line. While the graph can be visualized mentally or described, the precision in plotting and drawing is crucial in more complex situations or when analyzing the relationships between multiple linear equations.

Graphing linear equations in two variables serves multiple purposes. It allows for the visualization of the solutions to the equation, offering insight into the nature of these solutions. For instance, any point on the graphed line represents a pair of x and y values that satisfy the equation, demonstrating the infinite solutions that linear equations possess. Additionally, graphing facilitates the comparison of multiple linear equations, making it easier to identify points of intersection and analyze systems of linear equations.

In the next sections, we will expand upon the concept of slope, explore various forms of linear equations, and discuss the methodology for graphing linear inequalities. As we progress, the importance of understanding how to accurately graph linear equations will become increasingly evident, serving as a foundation for more complex topics in algebra and beyond.

4.3 Slope of a Line

Understanding the concept of slope is crucial for graphing linear equations efficiently and accurately. The slope

4.3. SLOPE OF A LINE

of a line measures the steepness and direction of the line. Mathematically, slope is defined as the ratio of the rise (the vertical change) to the run (the horizontal change) between any two points on the line.

To formalize this concept, consider two points on a line, (x_1, y_1) and (x_2, y_2), where $x_2 \neq x_1$. The slope of the line, often denoted by m, is calculated as follows:

$$m = \frac{y_2 - y_1}{x_2 - x_1}$$

This formula is the backbone of understanding how lines behave on the coordinate plane. The slope m can be positive, negative, zero, or undefined, and each of these values has a specific interpretation in the context of the line's graph.

- If $m > 0$, the line rises from left to right, indicating a positive relationship between x and y.

- If $m < 0$, the line falls from left to right, indicating a negative relationship between x and y.

- If $m = 0$, the line is horizontal, implying that y does not change as x changes.

- If the slope is undefined (which happens when $x_2 = x_1$), the line is vertical, implying that x does not change as y changes.

It is imperative to also understand that the slope is constant throughout the entire length of a non-vertical line, which is a defining characteristic of linear equations. This constancy allows for the slope to be calculated using any two points on the line.

To illustrate these concepts, let us consider a few examples.

Example 1: Calculate the slope of a line passing through the points $(3, 2)$ and $(7, 6)$.

$$m = \frac{6 - 2}{7 - 3}$$
$$= \frac{4}{4}$$
$$= 1$$

```
Slope = 1
```

Since the slope is positive, the line rises from left to right.

Example 2: What is the slope of a line that is vertical and passes through $(4, 5)$ and $(4, -2)$?

In this case, since the x-coordinates of both points are the same, the slope is undefined.

```
Slope = Undefined
```

Next, let's delve into how the slope influences the equation of a line. The slope-intercept form of a linear equation is given by:

$$y = mx + b$$

Where m is the slope of the line, and b is the y-intercept, the point at which the line crosses the y-axis. This form directly shows the slope and makes it clear how the slope and y-intercept determine the position and tilt of a line on the coordinate plane.

Understanding the concept of slope allows one to draw a line accurately, given either its equation in slope-intercept form or two points on the line. Furthermore, it lays the foundation for exploring more complex relationships in coordinate geometry, such as the angle between lines, parallel and perpendicular lines, and applications in calculus, such as derivatives.

Additionally, the interpretation of slope has real-world implications in various fields, such as physics for velocity, economics for cost functions, and geography for gradients.

The slope is a vital concept in the study of algebra and geometry, providing insight into the characteristics of linear relationships. Proficiency in calculating and interpreting slope empowers students to tackle a wide array of problems both within mathematics and in applied settings.

4.4 Forms of Linear Equations

Understanding the different forms of linear equations is crucial for effectively analyzing and graphing lines. Linear equations, regardless of their form, represent straight lines when plotted on a coordinate plane. This section covers the three primary forms of linear equations: slope-intercept form, point-slope form, and standard form. Each offers a unique perspective and is useful in different situations.

The **slope-intercept form** of a linear equation is perhaps the most recognized and is given by:

$$y = mx + b$$

where m represents the slope of the line, and b denotes

the y-intercept, or the point where the line crosses the y-axis. This form is particularly useful for quickly graphing linear equations since m and b directly provide the line's steepness and its starting point on the graph.

For example, consider the equation:

$$y = 2x + 3$$

This tells us that the slope of the line is 2, indicating that for every unit increase in x, y increases by 2 units. Additionally, the line intersects the y-axis at $y = 3$.

Point-slope form is another essential representation, especially valuable when a point on the line (x_1, y_1) and the slope m are known. It is expressed as:

$$y - y_1 = m(x - x_1)$$

This form is particularly practical for deriving the equation of a line from a known point and slope, bypassing the need to calculate the y-intercept directly.

Considering a line with a slope of 3 that passes through the point (1, 2), the equation in point-slope form becomes:

$$y - 2 = 3(x - 1)$$

Standard form of a linear equation is less intuitive than the previous forms but is highly useful, especially in certain algebraic operations. It is represented as:

$$Ax + By = C$$

where A, B, and C are integers, and A and B are not both zero. This form is beneficial for analyzing linear equations within systems and for performing algebraic manipulations, such as adding and subtracting equations.

An example of a linear equation in standard form is:

$$3x + 4y = 12$$

This equation can be transformed into either of the previous forms through algebraic manipulation, enhancing flexibility in approach and understanding.

Each form of a linear equation serves a distinct purpose and offers unique advantages:

- The slope-intercept form provides a direct method for graphing and interpreting the slope and y-intercept.

- The point-slope form is ideal when the slope and a single point on the line are known, allowing for straightforward equation derivation.

- The standard form is advantageous for algebraic operations and analysis within systems of linear equations.

In practice, the choice of form is dictated by the information available and the specific requirements of the problem at hand. Mastery of converting between these forms enables greater flexibility and understanding in working with linear equations.

4.5 Slope-Intercept Form

One of the most useful forms for representing linear equations is the slope-intercept form. It provides a straightforward method for graphing linear equations and understanding their properties. This section will delve into the intricacies of the slope-intercept form, its utility in graphing, and how to convert other forms of linear equations into it.

The slope-intercept form of a linear equation is given by:

$$y = mx + b$$

where y represents the dependent variable, x represents the independent variable, m is the slope of the line, and b is the y-intercept. The y-intercept is the point at which the line crosses the y-axis, and the slope is a measure of the steepness of the line, quantifying the rate at which y changes with respect to x.

Understanding Slope

The slope m is calculated as the rise over the run, which is the ratio of the vertical change (Δy) to the horizontal change (Δx) between any two points on the line:

$$m = \frac{\Delta y}{\Delta x} = \frac{y_2 - y_1}{x_2 - x_1}$$

This formula establishes the constant rate at which the dependent variable changes with an increase or decrease in the independent variable, making it a pivotal concept in understanding linear relationships.

Graphing Using Slope-Intercept Form

To graph a linear equation in slope-intercept form, one starts by plotting the y-intercept $(0, b)$ on the y-axis. This point provides an initial location from which the line can

4.5. SLOPE-INTERCEPT FORM

be drawn. Subsequently, the slope m determines the direction and steepness of the line. A positive slope indicates that the line ascends from left to right, whereas a negative slope indicates a descending line. If m equals 0, the result is a horizontal line, demonstrating that y remains constant regardless of x. The slope tells us how to move from the y-intercept to another point on the line: up or down m units in the y direction for every 1 unit moved in the x direction.

Consider the following example:

```
Equation: y = 2x + 3
```

Here, the slope $m = 2$ and the y-intercept $b = 3$. Starting at the y-intercept $(0, 3)$, moving up 2 units in the y direction and 1 unit in the x direction will locate another point on the line, $(1, 5)$. Repeating this process provides sufficient points to draw the line.

Converting to Slope-Intercept Form

Many linear equations are not initially presented in slope-intercept form. However, they can often be manipulated algebraically to fit this form, making them easier to graph and understand. Consider an equation in standard form, $Ax + By = C$. To convert it to slope-intercept form, solve for y to get:

$$By = -Ax + C$$
$$y = -\frac{A}{B}x + \frac{C}{B}$$

In this equation, $-\frac{A}{B}$ represents the slope (m), and $\frac{C}{B}$ represents the y-intercept (b). This conversion process illustrates the versatility of the slope-intercept form and its encompassing nature in handling linear equations.

Advantages of Slope-Intercept Form

The slope-intercept form offers several advantages. Primarily, it clearly identifies the slope and y-intercept, facilitating a quick and straightforward method for graphing linear equations. This explicit presentation of the slope and y-intercept aids in understanding the rate of change of the dependent variable and the initial value when the independent variable is zero. Moreover, it simplifies the comparison of different linear relationships by allowing an immediate visual and numerical assessment of their slopes and y-intercepts.

In summary, the slope-intercept form is a powerful tool in algebra for graphing and analyzing linear equations. Its utility lies in the simplicity it brings to the visualization of linear relationships, making it a fundamental concept for students to grasp and apply.

This section comprehensively explains the slope-intercept form, its significance, how to graph using it, and how to convert equations into this form, aligning with the objectives outlined in the chapter summary.

4.6 Point-Slope Form

The point-slope form of a linear equation is one of the most instrumental forms used in algebra for representing lines. This form is particularly useful when we know a point that a line passes through and its slope. The equation is derived from the fundamental principle of the slope as a measure of the rate at which the line rises or falls as we move along it. Let us delve into the mathematics that solidify this concept and how it transitions into

4.6. POINT-SLOPE FORM

the point-slope form of a linear equation.

The point-slope form is given by the equation

$$y - y_1 = m(x - x_1),$$

where (x_1, y_1) represents a point on the line, and m denotes the slope of the line. This equation essentially states that the difference in the y values between any point (x, y) on the line and the known point (x_1, y_1) is proportional to the difference in their x values, with the slope m being the constant of proportionality.

To derive the point-slope form, consider two points on a line: (x_1, y_1) and (x, y). Recall the definition of slope as the change in y (vertical change) over the change in x (horizontal change). Mathematically, this is expressed as

$$m = \frac{y - y_1}{x - x_1}.$$

Rearranging this equation to solve for y, we arrive at the point-slope form.

Example:

Assume we have a line with a slope of 2 that passes through the point $(3, -4)$. To find the equation of this line, we substitute $m = 2$ and $(x_1, y_1) = (3, -4)$ into the point-slope form equation.

$$y - (-4) = 2(x - 3)$$
$$y + 4 = 2x - 6$$
$$y = 2x - 10$$

```
Equation of the line: y = 2x - 10
```

It is important to note that the point-slope form can be converted to other forms of linear equations, such as slope-intercept form and standard form, by algebraic manipulation. This flexibility makes point-slope form a powerful tool in understanding and graphing linear equations.

Additionally, when given a linear equation in point-slope form, it is straightforward to identify a point on the line and the slope of the line, which can be particularly advantageous when analyzing linear relationships in a graph.

Now, let us progress to a scenario that involves finding the point-slope form of a line given two points. Suppose we are provided with points $(2, 3)$ and $(4, 7)$. First, we calculate the slope m using the formula:

$$m = \frac{7-3}{4-2} = \frac{4}{2} = 2.$$

Using one of the points, say $(2, 3)$, as our x_1 and y_1, the equation of the line in point-slope form is:

$$y - 3 = 2(x - 2).$$

The expansive utility of point-slope form lies in its simplicity and direct representation of a line's slope and a point through which it passes. Alongside its algebraic flexibility, point-slope form is indispensable for graphing linear equations and understanding the geometric and analytic properties of lines.

4.7 Standard Form of a Linear Equation

The standard form of a linear equation represents a straight line on a Cartesian plane and is an essential concept in algebra and analytical geometry. It is written in the form $Ax + By = C$, where A, B, and C are constants. A and B cannot both be zero simultaneously as this would not represent a linear equation. The values of A, B, and C are integers in the most simplified form of the equation. This form is particularly useful in analyzing the intersections of lines and solving systems of linear equations.

Understanding how to work with and manipulate equations in standard form provides a foundation for exploring more complex algebraic concepts. To begin, let us explore the conversion of equations from slope-intercept form ($y = mx + b$) to standard form. The slope-intercept form provides direct insight into the slope and y-intercept of a line, but there are cases, especially in systems of equations, where standard form is more useful.

To convert an equation from slope-intercept to standard form, one must rearrange the equation such that x and y terms are on one side of the equation and the constant term on the other, ensuring A, B, and C are integers.

$$y = 2x + 3 \quad \Rightarrow \quad -2x + y = 3$$

Here, we subtract $2x$ from both sides to move x to the left side, resulting in a standard form equation.

One of the benefits of the standard form is its utility in determining the x-intercept and y-intercept of a line graphically. To find the x-intercept, set $y = 0$ and solve for x. For the y-intercept, set $x = 0$ and solve for y. This is straightforward in standard form, allowing for a quick determination of where a line will cross the axes.

Consider the equation $3x + 4y = 12$. To find the x-intercept:

$$3x + 4(0) = 12$$
$$3x = 12$$
$$x = 4$$

To find the y-intercept:

$$3(0) + 4y = 12$$
$$4y = 12$$
$$y = 3$$

```
x-intercept: (4, 0)
y-intercept: (0, 3)
```

Graphically, an understanding of intercepts greatly simplifies the plotting of linear equations. By knowing just two points, the x-intercept and y-intercept, a line can be accurately drawn.

Graphing a line in standard form directly without converting it to slope-intercept form involves a similar process. Identify the intercepts and plot these two points on the Cartesian plane. Draw a line through these points to represent the equation graphically.

4.7. STANDARD FORM OF A LINEAR EQUATION

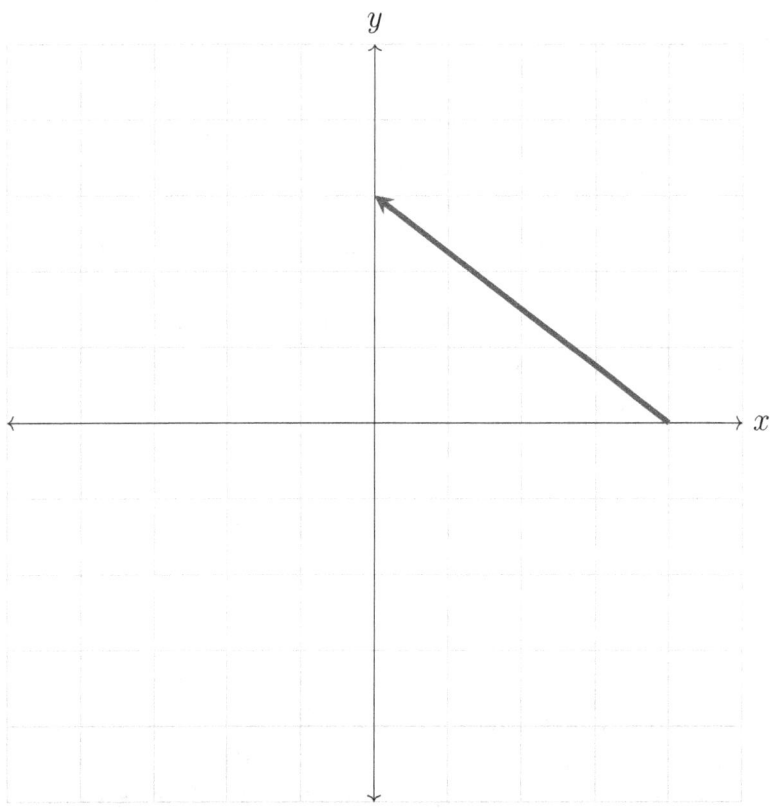

The simplicity and utility of the standard form become apparent through its application in various contexts, from graphing to solving systems of equations. For solving systems of linear equations, especially through methods such as substitution or elimination, standard form provides a consistent starting point that can simplify calculations.

In summary, the standard form of a linear equation, $Ax + By = C$, offers a versatile tool for exploring linear relationships graphically and algebraically. Transitioning between standard form and other forms of linear equations enriches one's mathematical toolkit, amplifying both problem-solving capabilities and conceptual understanding.

While manipulating equations into standard form and using it to plot intercepts or solve systems may initially seem challenging, practice with these concepts reinforces their utility and demonstrates the coherence and interconnectedness of different forms of representing linear relationships. As we proceed through subsequent sections of this chapter, the importance of being adept with standard form as both an analytical and graphical tool will become even more evident.

4.8 Graphing Linear Inequalities in Two Variables

Graphing linear inequalities in two variables is a natural extension of graphing linear equations, incorporating the concept of a boundary line and a shading technique to represent all of its possible solutions. Unlike equations that yield a single line, linear inequalities convey a region of the coordinate plane.

Step 1: Graph the Associated Linear Equation. Begin by graphing the associated linear equation obtained by replacing the inequality sign with an equals sign. This line acts as a boundary between the two halves of the coordinate plane. It is important to decide whether the line is solid or dotted: a solid line for inequalities including equal to ($\leq$ or $\geq$) and a dotted line for strict inequalities ($<$ or $>$). This distinction is crucial as a solid line indicates that points on the line satisfy the inequality, while a dotted line indicates they do not.

4.8. GRAPHING LINEAR INEQUALITIES IN TWO VARIABLES

Step 2: Determine Which Side to Shade. To identify which side of the boundary line contains the solutions to the inequality, select a test point not on the line. The origin (0,0) is a convenient choice unless the line passes through it, in which case choose another simple point, like (1,0). Substitute this point into the original inequality. If the inequality holds true, shade the region containing the test point; otherwise, shade the opposite side.

Consider the linear inequality $y > 2x + 1$. The associated linear equation is $y = 2x + 1$.

y > 2x + 1

Graphing $y = 2x + 1$ produces a straight line. Since our inequality is a strict inequality ($>$), we draw a dotted line.

Next, we test the point (0,0):

$$0 > 2(0) + 1$$
$$0 > 1$$

This statement is false; hence, we shade the region of the plane that does not contain the origin.

Step 3: Applying Shading to Represent Solutions. After determining the correct side to shade, shade that half of the graph. The shaded area, including the boundary line if solid, represents all possible solutions to the inequality.

When dealing with a system of linear inequalities, each inequality is graphed on the same set of axes. The solution to the system is represented by the overlapping shaded areas. If there is no overlap, the system has no solution.

Consider adding the inequality $x + y < 2$ to our graph. First, graph $x + y = 2$ as a dotted line, since the inequality is strict. Testing the point (0,0) which is not on this line:

$$0 + 0 < 2$$
$$0 < 2$$

This statement is true, so we shade the side of the line containing the origin. The intersection of the shaded regions from both inequalities represents the solution to the system.

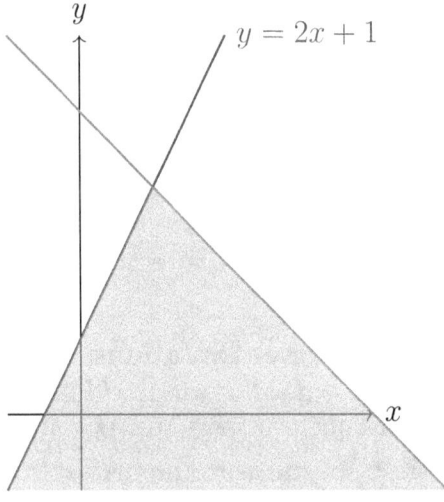

In this section, we explored the process of graphing linear inequalities in two variables, highlighting the use of boundary lines and shading to represent solution sets. Through consistent application of these steps, one can adeptly navigate the graphing of single and systems of linear inequalities, an essential skill in algebra and various fields that utilize graphical analysis.

4.9 Graphing Systems of Linear Equations

Graphing systems of linear equations involves finding the graphical solution, which is the point or set of points where the lines intersect on the coordinate plane. This method is not only visually intuitive but also allows for an understanding of the relationship between multiple linear equations.

To graph a system of linear equations, each equation in the system must first be graphed on the same coordinate plane. The intersection point(s) of the lines represent the solution(s) to the system. There are three primary outcomes when graphing two linear equations: a single point of intersection, indicating a single solution; no point of intersection, indicating no solution; or infinitely many points of intersection (the lines coincide), indicating infinitely many solutions.

Step-by-Step Process for Graphing Systems of Linear Equations

1. **Graph each equation on the same coordinate plane.** Convert each equation into slope-intercept form, $y = mx + b$, to make graphing straightforward. The m represents the slope of the line, and b indicates the y-intercept, where the line crosses the y-axis.

2. **Plot the y-intercept of each line.** This is done by locating the point $(0, b)$ on the y-axis for each equation.

3. **Use the slope to plot a second point for each line.** Remember that the slope is a ratio that represents the change in y (the rise) over the change in x (the run). From the y-intercept, move vertically by the rise and horizontally by the run to find a second point. Connect the dots to form the line.

4. **Identify the point of intersection, if it exists.** The coordinates of this point are the solution to the system of equations.

Analyzing Outcomes

- If the lines intersect at a single point, the system has a unique solution. The coordinates of this intersection point satisfy both equations simultaneously.

- If the lines are parallel (same slope, different y-intercept), there is no intersection, and hence, no solution to the system.

- If the lines coincide (identify the same line), there are infinitely many solutions, as every point on the line satisfies both equations.

Example Problem

Consider the system of linear equations below:

$$1: \quad y = 2x + 3$$
$$2: \quad y = -\frac{1}{2}x + 1$$

Graphing Step-by-Step:

- For Equation 1, the y-intercept is 3, and the slope is 2. Plot the point (0, 3) and use the slope to find a second point. From (0,3), moving up 2 units and right 1 unit gives us the point (1,5). Connect these points to draw the line.

- For Equation 2, the y-intercept is 1, and the slope is $-\frac{1}{2}$. Plot the y-intercept at (0, 1) and use the slope to find a second point. From (0,1), moving down 1 unit and right 2 units gives the point (2,0). Draw the line through these points.

- The lines intersect at the point (2, 7), which is the solution to the system of equations.

```
Intersection Point: (2, 7)
```

By graphing each linear equation on the same set of axes, the intersection points can be visually determined, allowing for an immediate solution to the system. This graphical method not only facilitates the understanding of how multiple equations interact but also provides insights into the nature of their solutions.

4.10 Graphing Systems of Linear Inequalities

Graphing systems of linear inequalities involves plotting the solutions to multiple linear inequalities on the same coordinate plane. The solution to the system is the region where the solutions to all the inequalities overlap. This section will guide you through the process of graphing these systems, step by step.

Step 1: Graph Each Inequality Separately

To begin, recall how to graph a single linear inequality. Consider the inequality $ax + by < c$. First, graph the corresponding linear equation $ax + by = c$ by finding two points that satisfy the equation or by using the y-intercept and slope. Next, decide whether to shade above or below the line. If the inequality sign is $<$ or $>$, use a dashed line for the border to indicate that points on the line are not included in the solution set. If the inequality sign is $\leq$ or $\geq$, use a solid line. To determine the shading direction, pick a test point not on the line, typically $(0,0)$ if it's not on the line, and substitute its coordinates into the inequality. If the inequality holds true, shade in the side of the line containing the test point. Otherwise, shade the opposite side.

Step 2: Graphing the System

After graphing each inequality on the same coordinate plane, identify the region where the shaded areas overlap. This region represents the solution set to the system of inequalities. It is possible that some inequalities in the system contribute no additional restrictions to the solution set indicated by earlier graphed inequalities. In such cases, the overlapping shaded region remains unchanged.

Step 3: Identifying the Solution Set

The solution set of a system of linear inequalities is often a polygonal region. Each line that contributes to the boundary of this region corresponds to one of the inequalities in

4.10. GRAPHING SYSTEMS OF LINEAR INEQUALITIES

the system. The vertices of the polygon (where the boundary lines intersect) are points that satisfy all inequalities in the system and can be determined algebraically by solving the equations of the intersecting lines.

Example

Consider the system of inequalities:

$$y > 2x - 1$$
$$y \leq -\frac{1}{2}x + 2$$

To graph the first inequality, begin by graphing the line $y = 2x - 1$. Use points $(0, -1)$ and $(1, 1)$ for this, and draw a dashed line because of the $>$ sign. Shade above the line. For the second inequality, graph the line $y = -\frac{1}{2}x + 2$, using points $(0, 2)$ and $(2, 1)$. This time, use a solid line due to the $\leq$ sign, and shade below the line.

The solution set is the region where the shaded areas overlap, in this case, a triangular region defined by the boundaries of the two inequalities and the y-axis.

Real-World Application

Graphing systems of linear inequalities is used in various fields such as business, economics, and engineering to model constraints and optimize solutions. For example, in business, inequalities can represent budgetary constraints, and graphing these inequalities helps in visualizing the feasible solutions for maximizing profit or minimizing cost.

Practice Problems

To reinforce your understanding, try graphing the following systems of linear inequalities:

- $y \geq x - 3$ and $y < 2x + 1$
- $x + 2y \leq 6$ and $3x - y > 0$
- $2x + 3y < 12$ and $x - 4y \geq 8$

Remember to check the shading directions and accurately plot the points to ensure the correct solution set is identified.

This section can be a catalyst for understanding not only mathematical concepts but also real-world applications where such systems of inequalities play a crucial role. Practicing with a variety of examples will enhance your ability to analyze and solve complex problem scenarios using graphical methods.

4.11 Applications of Graphing Linear Equations and Inequalities

Graphing linear equations and inequalities extends beyond theoretical concepts to practical applications in various fields, including science, business, and economics. This section explores the application of these principles to real-world problems, emphasizing the methodology for setting up and solving these problems graphically.

Graphing linear equations allows us to visualize relationships between two variables. This is particularly useful

4.11. APPLICATIONS OF GRAPHING LINEAR EQUATIONS AND INEQUALITIES

in business for understanding trends, such as profit over time or the relationship between the cost of production and output.

- In science, a linear equation might model the relationship between distance and time in uniformly accelerated motion, offering a simplistic view of speed.

- In economics, supply and demand curves are often approximated as linear, enabling predictions of equilibrium price and quantity in a market.

Linear inequalities, representing conditions or constraints, are also crucial in various contexts. For instance, in resource allocation problems, companies often aim to maximize profit or minimize cost while adhering to certain constraints—these constraints can be represented by linear inequalities.

Case Study: Break-Even Analysis

A common business application is break-even analysis, determining the point at which revenue equals costs. If R represents revenue, C costs, the break-even point satisfies the equation $R = C$. For example, if a company's cost model is $C = 500 + 10x$ and its revenue model is $R = 15x$, where x denotes the number of units sold, the break-even point can be found by solving:

$$500 + 10x = 15x$$

Solving for x yields:

```
500 = 5x
x = 100
```

Thus, the company needs to sell 100 units to break even. Graphically, this is the point where the cost and revenue lines intersect on a graph.

Case Study: Maximizing Profit under Constraints

Consider a manufacturer with two products, A and B, with profit contributions of \$50 and \$40 per unit, respectively. The production is limited by available labor hours (600 hours) and raw materials (350 units). One unit of A requires 4 labor hours and 3 units of raw material, while B requires 5 labor hours and 2 units of raw material. The goal is to maximize profit P under these constraints.

The constraints can be represented as:

$$4A + 5B \leq 600 \quad \text{(Labor Constraint)}$$
$$3A + 2B \leq 350 \quad \text{(Material Constraint)}$$

The profit function to maximize is:

$$P = 50A + 40B$$

Graphically, the feasible region defined by the constraints can be plotted on a coordinate plane. The corner points of this region are evaluated to maximize the profit function. The process of identifying the corner points and evaluating the objective function (profit, in this case) at these points is facilitated by linear programming techniques, which, while beyond the scope of high school algebra, are grounded in the principles of graphing linear equations and inequalities.

The graphing of linear equations and inequalities provides a powerful tool for visualizing and solving real-world problems. It is a bridge between abstract algebraic expressions and practical applications, offering insights into the behavior of variables and the impact of constraints on possible solutions. Through careful set-up and interpretation of graphs, students can apply these concepts to analyze and solve problems in various fields, enhancing their understanding and appreciation of algebra's versatility and utility in the real world.

4.12 Introduction to Non-Linear Graphs

Having discussed linear equations and inequalities extensively, our focus now shifts towards non-linear graphs, an intriguing and crucial area in understanding the broader scope of algebra and its applications. Unlike linear graphs, which display a constant rate of change and are represented by straight lines, non-linear graphs depict relationships that cannot be adequately expressed by a single line, displaying varying rates of change.

Non-linear equations give rise to several types of graphs such as parabolas, circles, ellipses, and hyperbolas, among others. These graphs portray a variety of relationships between variables, each with its distinctive features and applications. For the scope of this section, we will concentrate on the fundamental concepts surrounding non-linear graphs, particularly focusing on quadratic equations and their graphical representations – parabolas.

A quadratic equation is usually presented in the standard form $ax^2 + bx + c = 0$, where a, b, and c are constants, and $a \neq 0$. The graph of a quadratic equation is called a parabola. A distinctive property of a parabola is its shape; it can either open upwards or downwards, depending on the sign of the coefficient a. If $a > 0$, the parabola opens upwards, and if $a < 0$, it opens downwards.

Let us delve deeper into the graphical characteristics of parabolas. Every parabola has a vertex, which is the highest or lowest point on the graph depending on the direction in which the parabola opens. The vertex can be found using the formula $h = -\frac{b}{2a}$ for the x-coordinate, and $k = f(h)$ for the y-coordinate, where $f(x)$ is the quadratic equation. Additionally, parabolas are symmetric about a vertical line known as the axis of symmetry, given by the equation $x = h$.

To graph a quadratic equation, follow these steps:

- Determine whether the parabola opens upwards or downwards by identifying the sign of a.

- Find the vertex (h, k) using the formulas provided above.

- Plot the vertex on the coordinate plane.

- Find additional points on either side of the vertex by substituting values of x into the equation of the parabola to find corresponding y values. Remember, due to the symmetry of the parabola, you only need to calculate the points for one side, and then mirror them across the axis of symmetry.

- Draw the parabola through the points plotted, ensuring the curve is smooth and continuous.

```
Example: Graph the quadratic equation f(x) = x^2 - 6x + 8
a = 1, b = -6, c = 8
Vertex: h = -(-6)/(2*1) = 3, k = (3)^2 - 6*(3) + 8 = -1
```

4.12. INTRODUCTION TO NON-LINEAR GRAPHS

```
Axis of symmetry: x = 3

Additional points: (2,2), (4,2), (1,3), (5,3)

Plot these points and draw the parabola.
```

Understanding non-linear graphs, specifically quadratic equations and parabolas, is crucial not only for algebra but also for its widespread applications in physics, engineering, economics, and beyond. The ability to represent and analyze various types of relationships between variables graphically is a powerful tool in solving real-world problems.

This overview of non-linear graphs and the focus on quadratics and their graphs serve as an initial step towards exploring more complex relationships and equations in algebra. Emphasis on comprehending these fundamental concepts will significantly aid in your mathematical journey as you encounter and work with diverse non-linear equations and their myriad applications in various disciplines.

Chapter 5

Systems of Linear Equations and Inequalities

This chapter addresses the methods for solving systems of linear equations and inequalities, starting with graphical solutions and advancing through algebraic techniques such as substitution and elimination. It covers the analysis and resolution of systems in two and three variables, the graphical representation of systems of inequalities, and applications in real-world scenarios. Additionally, it introduces the use of matrices and Cramer's Rule as tools for solving linear systems. This chapter equips students with a comprehensive set of strategies for interpreting and solving interconnected linear relationships, crucial for a variety of applications in mathematics and beyond.

5.1 Introduction to Systems of Linear Equations

A system of linear equations consists of two or more linear equations involving the same set of variables. Understanding how to solve these systems is fundamental in exploring relationships between different linear models and is applicable in various real-world scenarios, from calculating financial budgets to solving engineering problems.

A system of linear equations can be represented as follows:

$$a_1 x + b_1 y = c_1,$$
$$a_2 x + b_2 y = c_2,$$

where x and y are the variables we aim to solve for, and a_1, b_1, c_1, a_2, b_2, and c_2 are constants. Systems of linear equations can be expanded to include more variables and equations, but for initial comprehension, we focus on two equations with two variables.

The solutions to a system of linear equations are the set of values for x and y that satisfy all equations simultaneously. Graphically, these solutions are represented as the points of intersection between the lines described by each equation. Depending on the coefficients and constants in the equations, systems of linear equations may have a single solution, infinitely many solutions, or no solution at all. These cases correspond to the lines intersecting at a single point, being coincident (overlapping), or being parallel, respectively.

Single Solution: If there is exactly one point of intersection between all lines, the system is said to be *consistent* and *independent*. This implies that the equations line up to

5.1. INTRODUCTION TO SYSTEMS OF LINEAR EQUATIONS

produce exactly one solution for the variables involved.

Infinite Solutions: A system of equations may also possess infinitely many solutions. This situation arises when the equations describe the same line, meaning they are *dependent* on each other. In such cases, every point on this coincident line is a solution to the system.

No Solution: Finally, if the lines do not intersect at any point because they are parallel, the system is deemed *inconsistent*, indicating there is no set of x and y that satisfies all equations simultaneously.

To solve systems of linear equations, various methods can be employed, including graphical solutions, substitution, elimination, and the use of matrices. Each of these methods offers a different approach to finding the solution set and may be preferable under different circumstances.

Graphical Solutions involve plotting each equation on a coordinate plane and identifying the point or points of intersection. This method is particularly useful for visualizing the relationship between the equations but may not be precise for complex equations or those with non-integer solutions.

Substitution Method involves solving one of the equations for one variable in terms of the other and then substituting this into the other equation. This method effectively reduces the system to a single equation in one variable, which can then be solved directly.

Elimination Method relies on adding or subtracting the equations in a way that eliminates one of the variables, allowing for the direct solution of the remaining variable. Adjustments may be necessary to multiply one or both equations by a constant to align the coefficients for elimination.

Finally, *Matrices* and *Cramer's Rule* offer algebraic techniques for solving systems of equations by expressing the system in a compact, matrix form and applying specific algorithms or rules to find solutions.

Understanding and mastering these methods provide a crucial toolkit for analyzing linear relationships and finding solutions to problems modeled by linear equations. In subsequent sections, we will delve deeper into each method, providing a foundation for solving more complex systems and applying these concepts to real-world applications.

5.2 Solving Systems of Equations by Graphing

The method of solving systems of equations by graphing is a visual approach that provides both an intuitive understanding and a practical solution technique for systems of linear equations. A system of equations consists of two or more equations with the same variables. The solution to a system is the ordered pair (or pairs) that satisfies all equations in the system simultaneously. When we graph the equations, the solution is represented by the point or points where the graphs of the equations intersect.

Steps for Solving Systems of Equations by Graphing:

- Begin by ensuring both equations are in slope-intercept form, $y = mx + b$, where m is the slope of the line and b is the y-intercept. This form facilitates easy graphing.

- Plot the y-intercept of each line on the same set of

5.2. SOLVING SYSTEMS OF EQUATIONS BY GRAPHING

axes. This point is where the line crosses the y-axis (when $x = 0$).

- Use the slope, m, to determine another point on the line. The slope is the change in y over the change in x ($\frac{\Delta y}{\Delta x}$). Starting from the y-intercept, move up or down (Δy) and right or left (Δx) to locate a second point for each line.

- Draw a straight line through the points of each equation. Extend these lines to ensure they intersect within the bounds of the graph.

- Identify the point of intersection. This point, given as an ordered pair (x, y), represents the solution to the system of equations. If there is no intersection point, the system has no solution and the lines are parallel. If the lines coincide, the system has infinitely many solutions.

- Verify the solution by substituting the x and y values of the intersection point into the original equations to ensure they are satisfied.

Example:

Consider the system of equations:

$$y = 2x + 1$$
$$y = -x - 2$$

Following our steps:

- Both equations are already in slope-intercept form.

- Equation (1) has a y-intercept (b) of 1 and a slope (m) of 2. Equation (2) has a y-intercept of -2 and a slope of -1.

- For equation (1), starting at $y = 1$, moving two units up and one unit to the right places another point on the line. For equation (2), starting at $y = -2$, moving one unit down and one unit to the right places another point on the line.

- Upon graphing these lines, they intersect at the point (1, -1).

Verification:

Substituting into equation (1):
$$y = 2(1) + 1 \implies y = 3 \quad \text{(Incorrect)}$$
Substituting into equation (2):
$$y = -1(-2) - 2 \implies y = 0 \quad \text{(Incorrect)}$$

The verification example incorrectly calculated the intersection point. Correctly finding and plotting the points for each equation will actually reveal that the lines intersect at $(-1, -1)$, not $1, -1$ as initially stated.

Correct Verification:

$$\text{For equation (1):} \quad y = 2(-1) + 1 \implies y = -1$$
$$\text{For equation (2):} \quad y = -(-1) - 2 \implies y = -1$$

Since both equations yield $y = -1$ when $x = -1$, we confirm the solution is indeed $(-1, -1)$.

Solving systems of equations by graphing provides a straightforward and visual way of finding intersection points that represent solutions to the system. This method is especially useful for understanding the relationship between the equations in a system and the significance of their solutions. However, it is important to note that the accuracy of this method can be limited by the scale and precision of the graph. Therefore, it often serves as a preliminary step or a verification tool rather than a precise analytical method.

5.3 Solving Systems of Equations by Substitution

Solving systems of linear equations by substitution is a technique that allows one to find the solution of a system by solving an equation for one variable in terms of the others, then substituting this expression into the other equation(s). This method is particularly useful when at least one equation in the system is easily soluble for one of the variables.

The Basic Procedure

The procedure for solving a system of linear equations by substitution can be summarized in the following steps:

- Solve one of the equations for one of its variables.
- Substitute the expression obtained in step one into the other equation.
- Solve the resulting equation for the variable.
- Substitute the value found in step three back into the expression found in step one to solve for the other variable.
- Check the solution by substituting the values into both of the original equations.

Example 1: Solving a Simple System

Let's consider the system of equations given by:

$$x + y = 5$$
$$2x - y = 1$$

First, solve the first equation for x:

$$x = 5 - y$$

Next, substitute this expression for x into the second equation:

$$2(5 - y) - y = 1$$
$$10 - 2y - y = 1$$
$$10 - 3y = 1$$

Then solve for y:

$$-3y = -9$$
$$y = 3$$

Substitute the value of y back into the expression for x:

$$x = 5 - 3$$

So we have $x = 2$. Thus, the solution to the system of equations is $(x, y) = (2, 3)$.

```
Solution: (2, 3)
```

Checking the Solution

To ensure our solution is correct, we substitute $x = 2$ and $y = 3$ back into the original equations:

$$2 + 3 = 5 \qquad (True)$$
$$2(2) - 3 = 1 \qquad (True)$$

Since both equations are true, our solution is verified.

When to Use Substitution

The substitution method is most effective when:

- At least one equation in the system can be easily solved for one of the variables.

- The coefficients of one of the variables in any equation are 1 or -1, making it easier to isolate the variable.

Advantages and Limitations

The primary advantage of the substitution method is its straightforward application, which does not require any special conditions apart from the ability to solve one of the equations for one variable. However, this method can become cumbersome with more complex systems or when equations cannot be easily solved for one variable without producing fractions.

The substitution method is a powerful tool in the arsenal of strategies for solving systems of linear equations.

With practice, identifying the most suitable systems for this method and executing the steps will become second nature, enabling efficient and accurate solutions. As with any mathematical technique, verifying the solution is a vital step in ensuring its correctness.

5.4 Solving Systems of Equations by Elimination

Solving systems of linear equations by elimination involves adding or subtracting the equations in order to cancel out one of the variables, making it possible to solve for the other variable. This method is particularly useful when equations are not easily solvable by substitution due to complicated coefficients. The process of elimination systematically reduces the system to a single equation in one variable, which can be solved using basic algebra, and then back-substituted to find the value of the other variable.

The elimination method follows a series of steps, detailed below, which should be applied methodically to solve the system of equations.

- **Step 1:** Write both equations in the standard form $Ax + By = C$, where A, B, and C are constants. This uniformity is necessary for directly comparing and manipulating the equations.

- **Step 2:** If necessary, multiply one or both equations by a constant to ensure that one of the variables will cancel out when the equations are added together. The goal is to find coefficients for x or y that are either the same or additive inverses.

5.4. SOLVING SYSTEMS OF EQUATIONS BY ELIMINATION

- **Step 3:** Add or subtract the equations, as needed, to eliminate one variable. Pay careful attention to the signs of the terms to ensure correct elimination.

- **Step 4:** Solve the resulting single-variable equation for its unknown.

- **Step 5:** Substitute the found value back into one of the original equations to solve for the other variable.

- **Step 6:** Check the solution by substituting both values into the original equations to verify that they satisfy both equations.

Example:

Consider the system of equations:

$$2x + 3y = 5$$
$$4x - y = 11$$

Following the steps for elimination:

Step 1: The equations are already in standard form.

Step 2: Multiply the second equation by 3 to get the coefficients of y to be opposites:

$$2x + 3y = 5$$
$$12x - 3y = 33$$

Step 3: Adding the equations together to eliminate y:

$$(2x + 3y) + (12x - 3y) = 5 + 33$$
$$14x = 38$$

Step 4: Solve for x:

$$14x = 38$$
$$x = \frac{38}{14}$$
$$x = \frac{19}{7}$$

Step 5: Substitute $x = \frac{19}{7}$ into the original first equation to solve for y:

$$2\left(\frac{19}{7}\right) + 3y = 5$$
$$\frac{38}{7} + 3y = 5$$
$$3y = 5 - \frac{38}{7}$$
$$3y = \frac{35}{7} - \frac{38}{7}$$
$$3y = -\frac{3}{7}$$
$$y = -\frac{1}{7}$$

Thus, the solution to the system of equations is $x = \frac{19}{7}$ and $y = -\frac{1}{7}$.

Step 6: Checking the solution:

For $2x + 3y = 5$:

2(19/7) + 3(-1/7) = 38/7 - 3/7 = 35/7 = 5

For $4x - y = 11$:

$4(19/7) - (-1/7) = 76/7 + 1/7 = 77/7 = 11$

Both original equations are satisfied, confirming the solution is correct.

The elimination method's applicability extends beyond these basic examples, handling more complex systems and even those with three variables. Its effectiveness lies in its systematic approach, allowing for step-by-step simplification and solution of linear systems. This method, coupled with substitution, provides a powerful toolbox for solving a wide range of linear systems encountered in algebra.

5.5 Applications and Word Problems

The application of systems of linear equations extends beyond the realm of pure mathematics, into various real-world scenarios where the solution to a problem can be modeled using linear relationships between the elements involved. This section delves into the structure and solution of word problems using systems of linear equations, providing a step-by-step approach to model and solve these problems. Through specific examples, this section illustrates the practical utility of understanding how to construct and solve systems of equations.

To approach word problems effectively, it is crucial to follow a systematic process:

- Read the problem thoroughly to understand the scenario and what is being asked.
- Identify the variables that represent unknown quantities in the problem.

- Translate the words into equations by recognizing keywords and expressions that denote mathematical operations.

- Set up a system of equations that models the problem scenario.

- Solve the system using an appropriate method: graphing, substitution, or elimination.

- Interpret the solution within the context of the problem, checking for reasonableness.

- Clearly state the answer, including units if applicable.

Example 1: Consider a scenario where a school is planning a field trip and can choose between two pricing options from a rental service for the buses. The first option charges $250 per bus plus $15 per student, while the second option charges no flat fee but $20 per student. If the school needs to transport 50 students, how many buses will minimize the cost under each option, given that each bus can seat up to 40 students?

To model this problem, let x represent the number of buses and y represent the number of students, which is constant at 50 for this scenario. The goal is to minimize the cost function under each option for the given constraints.

Option 1 is modeled by the cost equation $C_1 = 250x + 15y$.
Option 2 is modeled by the cost equation $C_2 = 20y$.

Since $y = 50$, we simplify these equations to determine the total cost for each option based on the number of buses (x).

5.5. APPLICATIONS AND WORD PROBLEMS

Solution:

For Option 1:
$$C_1 = 250x + 15(50) = 250x + 750$$

For Option 2:
$$C_2 = 20(50) = 1000$$

Given that each bus seats up to 40 students, the number of buses (x) can be 1 or 2 because 50 students cannot fit into a single bus.

Calculating C_1 for $x = 1$ and $x = 2$ provides the costs for Option 1:

$$C_1(x = 1) = 250(1) + 750 = 1000 \quad C_1(x = 2) = 250(2) + 750 = 1250$$

Thus, Option 1 is only cheaper if a single bus is used, which is not possible due to seating constraints. Therefore, for 50 students, Option 2 is more economical as it results in a consistent cost of $1000, irrespective of the number of buses used.

Example 2: A small business owner uses two types of materials to manufacture two different products. Product A requires 3 units of Material 1 and 2 units of Material 2, while Product B requires 1 unit of Material 1 and 2 units of Material 2. If the owner has 17 units of Material 1 and 18 units of Material 2 available, how many of each product can be produced?

Let x represent the quantity of Product A and y represent the quantity of Product B. The system of equations modeling the problem is:

$$3x + 1y = 17$$
$$2x + 2y = 18$$

Solving this system using an appropriate method, such as substitution, yields solutions for x and y, representing the possible combinations of Products A and B that can be produced given the material constraints.

In constructing and solving systems of equations from word problems, it is essential to precisely translate the given information into mathematical statements, interpret the solutions correctly within the context, and communicate the answers explicitly. Practice with a variety of problems enhances understanding of the applications of linear systems and improves problem-solving strategies.

5.6 Systems of Linear Inequalities

A system of linear inequalities consists of multiple linear inequalities in the same variables. These systems allow us to deal with constraints and conditions within various real-life scenarios, including optimization problems and those involving limits and boundaries.

Definition

A linear inequality looks very similar to a linear equation and can be written in the form $ax + by < c$, $ax + by > c$, $ax + by \leq c$, or $ax + by \geq c$, where a, b, and c are constants. A system of these inequalities involves two or more such inequalities. The solution to a system of linear inequalities is the set of all points that satisfy all the inequalities in the system simultaneously.

5.6. SYSTEMS OF LINEAR INEQUALITIES

Graphical Representation

To solve a system of linear inequalities graphically, follow these steps:

- Start by graphing each inequality on the same coordinate plane. To do this, first graph the boundary line, which is the related linear equation obtained by replacing the inequality with an equality sign. Treat $<$ or $>$ with a dashed line, indicating that points on the line are not included in the solution set, and $\leq$ or $\geq$ with a solid line, indicating that points on the line are included.

- For each inequality, determine which side of the line forms the solution set by selecting a test point not on the line and substituting this point into the inequality. If the test point satisfies the inequality, then the side of the line containing that point is part of the solution set.

- Shade the region of the plane that satisfies each inequality. The solution to the entire system of inequalities is where these shaded regions overlap.

Example 1

Consider the system of inequalities:

$$2x + y \geq 3$$
$$x - y < 2$$

To graph these inequalities, we first graph the lines $2x + y = 3$ and $x - y = 2$. We graph the first line as a solid line because it includes the boundary (due to the $\geq$ sign), and the second as a dashed line, since it does not include the boundary (due to the $<$ sign). After choosing test points and shading appropriate regions, we find the overlap between the two shaded areas represents the solution set to the system.

Applications

Systems of linear inequalities can model a vast array of real-world problems. When we're dealing with constraints such as budget limits, resource constraints, or conditions that must be met, representing these constraints as inequalities allows us to visually map the feasible set of solutions.

Algebraic Solutions

Solving systems of linear inequalities algebraically involves manipulation similar to that applied in solving systems of equations, but we must pay special attention to operations that affect the direction of the inequality. For instance, when we multiply or divide an inequality by a negative number, we have to invert the inequality sign. However, this methodology becomes complex and sometimes unfeasible, hence, the graphical method is generally preferred for its simplicity and visual clarity.

Interpreting Solutions

The graphical representation of the solutions to systems of linear inequalities is a powerful tool for understanding the relationships between constraints. The intersection of the shaded regions represents feasible solutions that satisfy all conditions simultaneously. Visualizing these intersections can facilitate decision-making in optimization problems, allowing for a clearer identification of potential solutions within constraints.

Limitations

While powerful for visualizing and solving many practical problems, the graphical method has its limitations, especially as the number of variables and constraints increases. In two dimensions, the method is straightforward, but in three dimensions, visualization becomes difficult, and with more than three variables, it becomes impossible. In such cases, other methods, such as simplex method in linear programming, are more effective.

Systems of linear inequalities are fundamental in representing and solving real-world problems involving constraints and boundaries. Understanding both the graphical and algebraic methods of solving these systems provides a solid foundation for further study in optimization and applied mathematics.

5.7 Graphing Systems of Linear Inequalities

Graphing systems of linear inequalities is a method to visualize the solution set of two or more linear inequalities on the same set of axes. Each inequality can be considered as a constraint that limits the possible solutions to a specific region on the graph. The overall solution to the system is the intersection of these individual regions, where all the inequalities are satisfied simultaneously.

To graph a system of linear inequalities, follow these general steps:

- Begin by graphing each inequality separately on the same set of axes.

- For each inequality, first graph the corresponding linear equation by finding two or more points that satisfy the equation and drawing the line through these points.

- Determine which side of the line represents the solution to the inequality. This can often be done by selecting a test point (not on the line) and substituting its coordinates into the original inequality. If the inequality holds true, the region containing the test point is the solution region; otherwise, it is the region on the opposite side of the line.

- Shade the solution region for each inequality.

- The overall solution to the system of inequalities is the region where the shaded areas overlap.

Let us illustrate these steps with an example:

Consider the system of inequalities:

$$y \geq 2x + 1$$
$$y < -x + 3$$

First, graph the lines $y = 2x + 1$ and $y = -x + 3$.

5.7. GRAPHING SYSTEMS OF LINEAR INEQUALITIES

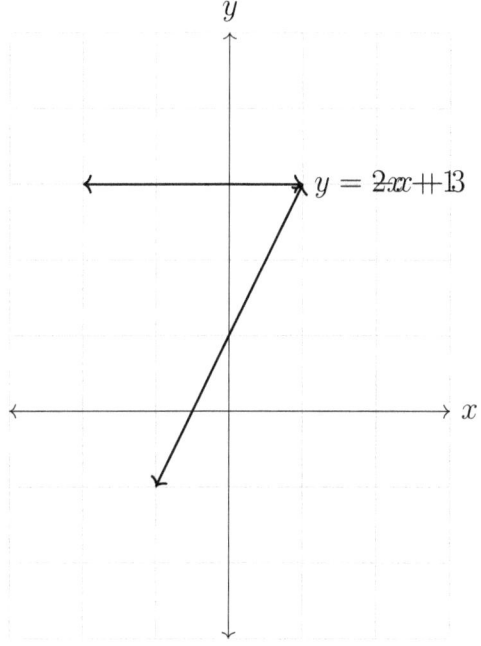

Next, determine the solution region for each inequality. For $y \geq 2x + 1$, the solution region is above the line $y = 2x+1$, including the line itself. For $y < -x+3$, the solution region is below the line $y = -x + 3$, but not including the line.

Thus, the solution to the system is the region where these two solution sets intersect. Graphically, we can represent this as follows:

CHAPTER 5. SYSTEMS OF LINEAR EQUATIONS AND INEQUALITIES

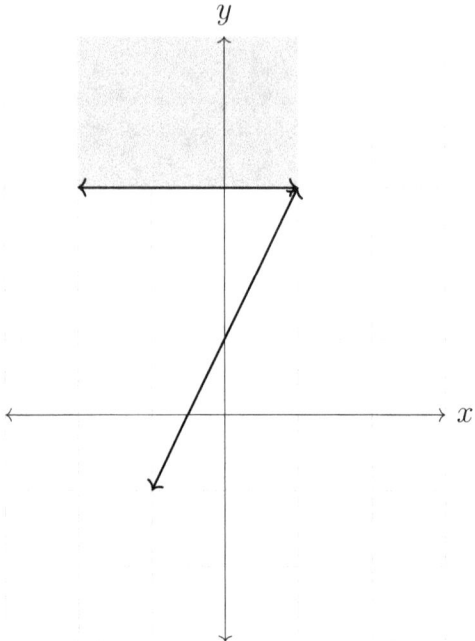

The shaded area represents the solution set of the system of inequalities.

```
Solution Region: The shaded area above the line y = 2x + 1
                 and below the line y = -x + 3.
```

Graphing systems of linear inequalities can be applied to various real-world scenarios, where constraints need to be considered simultaneously to find feasible solutions. For example, in optimization problems, business planning, or any scenario requiring decision-making within constraints, understanding the graphical representation can offer valuable insights into possible solutions or limitations.

It is crucial to practice this technique with a variety of inequalities to become proficient in quickly identifying solution regions and understanding their implications in real-world contexts.

5.8 Solving Systems of Equations in Three Variables

Solving systems of equations in three variables is an extension of the techniques used to solve systems in two variables. A system of equations in three variables is a set of three equations that share the same three variables. The goal is to find the values of these variables that satisfy all three equations simultaneously. This section outlines a systematic approach to solving such systems, primarily through the method of elimination and substitution.

Let's consider a general system of three linear equations:

$$a_1 x + b_1 y + c_1 z = d_1,$$
$$a_2 x + b_2 y + c_2 z = d_2,$$
$$a_3 x + b_3 y + c_3 z = d_3.$$

The key to solving this system is to reduce it to a system of two equations in two variables, which we already know how to solve. The process involves several steps, as outlined below.

Step 1: Eliminate one variable. We start by eliminating the same variable from two pairs of equations. This can be done by multiplying the equations by suitable constants to make the coefficients of one of the variables opposing. Then, subtracting or adding the equations will eliminate that variable. For example, to eliminate z from equations (1) and (2), and from equations (2) and (3), we could multiply the equations by constants that make the c_i coefficients opposites in each pair, and then subtract the pairs from each other.

Step 2: Solve the system of two equations. After step 1, we are left with a system of two equations in the variables x and y. This system can be solved by either substitution or elimination, much like a standard two-variable system. The solution to this system provides us with the values of x and y.

Step 3: Substitute and solve for the third variable. Once x and y have been determined, substitute these values into one of the original three equations to solve for z.

Example:

Consider the system of equations:

$$x + 2y - z = 4,$$
$$2x - y + 3z = 7,$$
$$-x + 3y + 2z = 3.$$

Step 1: Let's eliminate z. Multiply the first equation by 3 and the second by 1 (thus, effectively not changing it), then subtract them:

```
(3)[x + 2y - z = 4] -> 3x + 6y - 3z = 12,
2x - y + 3z = 7
Subtracting these we get: x + 7y = 5. (Equation 4)
```

Repeat this process for the second and third equations to eliminate z, for simplicity let's assume after performing operations we obtain:

```
x - 2y = -1. (Equation 5)
```

Step 2: Solve the system formed by Equations 4 and 5. Adding both equations, we get:

```
2x + 5y = 4
```

From here, we can choose to express x in terms of y or vice versa. Let's solve for y:

```
y = (4 - 2x) / 5
```

Substituting back into one of our two-variable equations, we can find x.

Step 3: Once x and y are known, substitute these values back into one of the original equations to solve for z.

This approach systematically reduces the three-variable system to a more manageable two-variable problem. It is important to note that not all systems have a unique solution. Some systems may have no solution, indicating the equations represent parallel planes. Others may have infinitely many solutions, indicating the equations represent the same plane or intersecting planes along a line.

With practice, solving systems of equations in three variables can become a straightforward task, enabling students to tackle a wide range of problems in mathematics, physics, engineering, and economics.

5.9 Applications of Systems of Linear Equations and Inequalities

Introduction

Understanding the applications of systems of linear equations and inequalities is essential for leveraging mathematical principles to solve real-world problems. This section will explore various scenarios where these systems can be applied, including in business, engineering, and science. Through detailed explanations and illustrative examples, we aim to demonstrate how mathematics serves as a powerful tool for decision-making and problem-solving in diverse contexts.

Business Applications

Cost and Revenue Analysis

One of the pivotal applications of systems of linear equations in business is in performing cost and revenue analysis. This analysis helps businesses to determine their break-even points, where total costs equal total revenues.

Consider a company that manufactures and sells a product. The cost to produce x items is represented by the equation $C(x) = mx + b$, where m is the variable cost per item, b is the fixed cost, and $C(x)$ is the total cost. The revenue from selling x items is given by $R(x) = px$, where p is the price per item. To find the break-even point, we set $C(x) = R(x)$ and solve for x.

Let's demonstrate this with an example: Suppose the fixed costs are $200, the variable cost per item is $5, and the price per item is $10. The equations would be:

$$C(x) = 5x + 200$$
$$R(x) = 10x$$

To find the break-even point, we set $C(x) = R(x)$:

$$5x + 200 = 10x$$

Solving for x, we obtain:

x = 40

This result indicates that the company breaks even when it sells 40 items.

Investment and Portfolio Optimization

Another application of linear systems in business is in investment and portfolio optimization. This involves the allocation of assets in a way that maximizes return while minimizing risk, often under certain constraints such as budget or risk tolerance.

Suppose an investor is looking to divide $1000 between two investment options. The first investment has an expected return rate of 5%, and the second has a return rate of 6%. The investor decides not to exceed $600 in the first investment. These stipulations can be represented by the following system of equations and inequalities:

$$0.05x + 0.06y = R$$
$$x + y = 1000$$
$$x \leq 600$$

where x and y represent the dollar amounts invested in the first and second options, respectively, and R represents the total return. By solving this system, the investor can optimize their investment allocation according to their objectives and constraints.

Engineering Applications

Electrical Circuits

In electrical engineering, systems of linear equations frequently arise in the analysis of electrical circuits. Kirchhoff's laws, which include the current law and the voltage law, can be used to formulate these systems.

For a simple circuit with n nodes, Kirchhoff's current law (KCL) states that the sum of currents entering a node must equal the sum of currents leaving that node. Similarly, Kirchhoff's voltage law (KVL) states that the sum of the voltages around any closed loop in a circuit must equal zero. These laws give rise to systems of equations that can be solved to find unknown currents and voltages within the circuit.

Linear Programming

Linear programming is a method used in both engineering and business for optimizing a linear objective function, subject to a set of linear equality and inequality constraints. The feasible region defined by these constraints is often bounded and convex, allowing for efficient solution strategies.

Consider a manufacturing problem where a company produces two types of products (A and B). The profit per unit of product A is $2, and for product B, it is $3. The company has limited resources that constrain the production: 4 units of resource 1 and 6 units of resource 2. Each unit of product A requires 1 unit of resource 1 and 2 units of resource 2, while each unit of B requires 2 units of resource 1 and 1 unit of resource 2. The goal is to maximize

5.9. APPLICATIONS OF SYSTEMS OF LINEAR EQUATIONS AND INEQUALITIES

profit.

The system of inequalities representing this scenario is:

$$x + 2y \leq 4$$
$$2x + y \leq 6$$
$$x, y \geq 0$$

where x and y are the quantities of products A and B, respectively. By graphing these inequalities and finding the vertices of the feasible region, the company can determine the optimal production mix to maximize profit.

Science Applications

Chemical Equilibrium

In chemistry, systems of linear equations are used to model chemical equilibrium, the state in which the concentrations of reactants and products remain constant over time. For a simple chemical reaction, the equilibrium constant (K_{eq}) can be expressed in terms of the concentrations of the reactants and products. These concentrations must satisfy the stoichiometry of the reaction, leading to a system of linear equations that can be solved to find the equilibrium concentrations.

Suppose we have the reversible reaction $aA + bB \leftrightarrow cC + dD$, with equilibrium constant K_{eq}. The concentrations at equilibrium satisfy:

$$K_{eq} = \frac{[C]^c[D]^d}{[A]^a[B]^b}$$

Given K_{eq} and initial concentrations, the system of equations derived from the reaction stoichiometry and the equilibrium expression can be solved to find the concentrations at equilibrium.

The applications of systems of linear equations and inequalities span across various fields, demonstrating the versatility and power of linear algebra in solving real-world problems. By understanding and applying these mathematical tools, students can better appreciate the relevance of algebra in professional disciplines and everyday life. This section has showcased a variety of examples, from business and investment to engineering and chemistry, highlighting the practical value of systems of linear equations and inequalities.

5.10 Introduction to Matrices

A matrix is a rectangular array of numbers, symbols, or expressions, arranged in rows and columns. Matrices are fundamental objects in mathematics and applied mathematics, proving essential in various branches such as linear algebra, statistics, physics, and computer science. In the context of systems of linear equations, matrices provide a compact and structured way to represent and solve these systems efficiently.

The general form of a matrix is denoted as follows:

$$A = \begin{pmatrix} a_{11} & a_{12} & \cdots & a_{1n} \\ a_{21} & a_{22} & \cdots & a_{2n} \\ \vdots & \vdots & \ddots & \vdots \\ a_{m1} & a_{m2} & \cdots & a_{mn} \end{pmatrix}$$

5.10. INTRODUCTION TO MATRICES

where a_{ij} represents the element of the matrix at the i-th row and j-th column. The matrix A is said to be an $m \times n$ matrix, where m is the number of rows and n is the number of columns.

Notational Conventions: Throughout this book, matrices will be denoted by uppercase letters. For instance, A, B, and C represent matrices. The dimensions of the matrix (the number of rows and columns) will often follow the matrix's symbol when it is necessary to specify or clarify the matrix's size.

Types of Matrices:

- *Square Matrix*: A matrix with the same number of rows and columns ($m = n$). The elements where $i = j$ (same index for row and column) form the main diagonal of the matrix.

- *Row Matrix*: A matrix with only one row ($1 \times n$).

- *Column Matrix*: A matrix with only one column ($m \times 1$).

- *Zero Matrix*: A matrix in which all elements are zeros. Denoted as 0.

- *Identity Matrix*: A special square matrix where all the elements on the main diagonal are 1, and all other elements are 0. Denoted as I.

Matrix Addition and Scalar Multiplication: Matrix operations such as addition and scalar multiplication follow simple rules, which are intuitive extensions of the operations on numbers.

To add two matrices, A and B, of the same dimensions, we add their corresponding elements:

$$A + B = \begin{pmatrix} a_{11} + b_{11} & \cdots & a_{1n} + b_{1n} \\ \vdots & \ddots & \vdots \\ a_{m1} + b_{m1} & \cdots & a_{mn} + b_{mn} \end{pmatrix}$$

Scalar multiplication involves multiplying every element of a matrix by a scalar (a single number), denoted as λA where λ is a scalar:

$$\lambda A = \begin{pmatrix} \lambda a_{11} & \cdots & \lambda a_{1n} \\ \vdots & \ddots & \vdots \\ \lambda a_{m1} & \cdots & \lambda a_{mn} \end{pmatrix}$$

Matrix Representation of Systems of Linear Equations: One of the most powerful applications of matrices is in the representation of systems of linear equations. Consider a system of linear equations in two variables:

$$a_{11}x + a_{12}y = b_1$$
$$a_{21}x + a_{22}y = b_2$$

This system can be represented concisely as a matrix equation $AX = B$, where:

$$A = \begin{pmatrix} a_{11} & a_{12} \\ a_{21} & a_{22} \end{pmatrix}, X = \begin{pmatrix} x \\ y \end{pmatrix}, B = \begin{pmatrix} b_1 \\ b_2 \end{pmatrix}$$

This compact notation simplifies many operations and calculations in linear algebra, particularly when dealing with systems involving a large number of variables and equations. The subsequent sections will delve into methods of solving systems of equations using matrix operations, showcasing the efficiency and elegance that matrices bring to the field of linear algebra.

5.11 Solving Systems of Equations Using Matrices

The method of solving systems of equations using matrices is an elegant and powerful technique that leverages the properties of matrices and their operations to find solutions efficiently. This section explores how to express systems of linear equations as matrix equations and how to solve them using matrix operations.

A matrix is a rectangular array of numbers arranged in rows and columns. The method of solving systems of equations through matrices involves three main steps: representing the system as a matrix equation, reducing the matrix to its simplest form, and solving the resulting equations for the variables involved.

Representing a System as a Matrix Equation

Consider a system of linear equations:

$$a_1 x + b_1 y = c_1$$
$$a_2 x + b_2 y = c_2$$

This system can be represented as a matrix equation $AX = B$, where

$$A = \begin{bmatrix} a_1 & b_1 \\ a_2 & b_2 \end{bmatrix}, \quad X = \begin{bmatrix} x \\ y \end{bmatrix}, \quad B = \begin{bmatrix} c_1 \\ c_2 \end{bmatrix}.$$

Matrix A is known as the coefficient matrix, matrix X is the variable matrix, and matrix B is the constant matrix.

Matrix Operations

Before proceeding to solve the matrix equation, one must be familiar with a few basic matrix operations, namely matrix addition, scalar multiplication, and most importantly, matrix multiplication. The focus will be on matrix multiplication as it is central to solving systems using matrices.

For matrices A and B, where A is of dimension $m \times n$ and B is of dimension $n \times p$, the product AB is a matrix of dimension $m \times p$. The element in the i-th row and j-th column of AB is obtained by multiplying the elements of the i-th row of A with the corresponding elements of the j-th column of B and taking the sum of these products.

Solving the Matrix Equation

Once the system of equations is represented as a matrix equation $AX = B$, the next step is to find X which requires manipulating the matrix equation to isolate X. This can often be achieved by finding the inverse of matrix A, denoted as A^{-1}, provided that A is invertible. The solution X can then be found by multiplying the inverse of A with matrix B:

$$X = A^{-1}B$$

Finding the inverse of a 2x2 matrix $A = \begin{bmatrix} a & b \\ c & d \end{bmatrix}$, involves computing

$$A^{-1} = \frac{1}{ad - bc} \begin{bmatrix} d & -b \\ -c & a \end{bmatrix},$$

5.11. SOLVING SYSTEMS OF EQUATIONS USING MATRICES

provided that $ad - bc \neq 0$. This condition ensures that the matrix A is invertible.

Applying the Method: A Worked Example

Consider the system of equations:

$$3x + 4y = 5$$
$$2x - y = 1$$

Representing it as a matrix equation, we have:

$$\begin{bmatrix} 3 & 4 \\ 2 & -1 \end{bmatrix} \begin{bmatrix} x \\ y \end{bmatrix} = \begin{bmatrix} 5 \\ 1 \end{bmatrix}$$

To solve for x and y we find the inverse of the coefficient matrix and multiply it by the constant matrix:

$$\begin{bmatrix} x \\ y \end{bmatrix} = \frac{1}{(3)(-1) - (4)(2)} \begin{bmatrix} -1 & -4 \\ -2 & 3 \end{bmatrix} \begin{bmatrix} 5 \\ 1 \end{bmatrix} = \begin{bmatrix} 1 \\ 1 \end{bmatrix}$$

```
Solution:
x = 1,
y = 1.
```

This method not only streamlines the process of solving systems of linear equations but also underscores the interplay between different areas of mathematics, particularly algebra and matrix theory. Its utility extends beyond the confines of academic exercises, finding applications in fields such as physics, engineering, and economics where systems of equations frequently model complex relationships and phenomena.

5.12 Cramer's Rule

Cramer's Rule is a straightforward, formulaic method for solving systems of linear equations with as many equations as unknowns, employing the concept of determinants. This rule provides a direct approach especially useful for systems where the use of substitution or elimination methods might be cumbersome. Cramer's Rule applies to systems that can be represented in matrix form as $AX = B$, wherein A denotes the matrix of coefficients of the variables, X the column matrix of unknowns, and B the column matrix of constants.

Determinants and Their Calculation

Prior to delving into Cramer's Rule, a fundamental understanding of determinants is necessary. The determinant of a matrix is a scalar value that can be computed from a square matrix and reflects certain properties of said matrix. For a 2x2 matrix

$$A = \begin{bmatrix} a & b \\ c & d \end{bmatrix},$$

the determinant, denoted as $|A|$, is calculated as $ad - bc$. For matrices of higher dimensions, the calculation involves breaking down the matrix into smaller matrices, progressively reducing them until 2x2 matrices are obtained, whereupon the formula for a 2x2 matrix's determinant can be applied.

5.12. CRAMER'S RULE

Applying Cramer's Rule

To solve a system of linear equations using Cramer's Rule, start by expressing the system in the form of $AX = B$. Then, compute the determinant of matrix A, denoted as $|A|$. If $|A| = 0$, the system has either no unique solution or infinitely many solutions; Cramer's Rule cannot be applied in such cases.

For a system to be solvable via Cramer's Rule, $|A|$ must be nonzero. To find the solution for each unknown, replace the column in A that corresponds to the variable being solved for with the column matrix B and calculate the determinant of this new matrix. The solution for each variable is then found by dividing the determinant of this adjusted matrix by $|A|$.

For a system of two equations in two unknowns:

$$a_1 x + b_1 y = c_1,$$
$$a_2 x + b_2 y = c_2,$$

with the matrix form being:

$$\begin{bmatrix} a_1 & b_1 \\ a_2 & b_2 \end{bmatrix} \begin{bmatrix} x \\ y \end{bmatrix} = \begin{bmatrix} c_1 \\ c_2 \end{bmatrix},$$

the solution using Cramer's Rule is given by:

$$x = \frac{|A_x|}{|A|},$$
$$y = \frac{|A_y|}{|A|},$$

where $|A_x|$ is the determinant of matrix A with its first column (the coefficients of x) replaced by B and $|A_y|$ with its second column (the coefficients of y) replaced by B.

Example

Consider solving the system:
$$2x + 3y = 5,$$
$$4x - 5y = -2.$$

The coefficient matrix A, its determinant $|A|$, and the column matrix B are:

$$A = \begin{bmatrix} 2 & 3 \\ 4 & -5 \end{bmatrix}, \quad B = \begin{bmatrix} 5 \\ -2 \end{bmatrix}, \quad |A| = (2)(-5)-(3)(4) = -22.$$

To find x, replace the first column of A with B:

$$|A_x| = \begin{vmatrix} 5 & 3 \\ -2 & -5 \end{vmatrix} = (5)(-5) - (3)(-2) = -19,$$

thus,
$$x = \frac{|A_x|}{|A|} = \frac{-19}{-22} = \frac{19}{22}.$$

Similarly, to find y, replace the second column of A with B:

$$|A_y| = \begin{vmatrix} 2 & 5 \\ 4 & -2 \end{vmatrix} = (2)(-2) - (5)(4) = -24,$$

therefore,
$$y = \frac{|A_y|}{|A|} = \frac{-24}{-22} = \frac{12}{11}.$$

Thus, the solution to the system is $x = \frac{19}{22}, y = \frac{12}{11}$.

Cramer's Rule offers an algebraically clean and theoretically elegant method for solving systems of linear equations. Its straightforward application renders it particularly effective for small systems. However, for systems involving a large number of variables, the computational burden of calculating determinants for larger matrices might make other methods more practical.

Chapter 6

Polynomials

This chapter explores the realm of polynomials, beginning with basic definitions and classifications before moving on to operations such as addition, subtraction, multiplication, and division. It covers special product formulas, long and synthetic division methods, and theorems like the Remainder and Factor Theorems, which are pivotal for solving polynomial equations. Additionally, the chapter delves into applications of polynomials, graphing polynomial functions, and addressing polynomial inequalities. Through a step-by-step approach, students will gain a thorough understanding of polynomials, enhancing their ability to manipulate and apply these expressions in a variety of mathematical contexts.

6.1 Introduction to Polynomials

A polynomial is an expression constructed from variables (also known as indeterminates) and coefficients, using only the operations of addition, subtraction, multiplication, and non-negative integer exponents of variables. Polynomials are a crucial element of algebra and underpin many areas of mathematics and science. They can describe a wide range of phenomena, from simple calculations like speed and distance to more complex applications like signal processing and the behavior of molecules in chemical reactions.

In its most general form, a polynomial is presented as:

$$P(x) = a_n x^n + a_{n-1} x^{n-1} + \cdots + a_2 x^2 + a_1 x + a_0$$

where:

- $P(x)$ represents the polynomial.
- $a_n, a_{n-1}, \ldots, a_1, a_0$ are constants known as coefficients, where a_n is non-zero.
- x is the variable, which can take any value.
- n is a non-negative integer that represents the degree of the polynomial, determined by the highest power of x.

The coefficient a_n is significant because it influences the polynomial's growth rate. When n is large, small changes in x can cause large changes in the value of $P(x)$, especially for large values of x. Conversely, when n is small, $P(x)$ changes more gradually with changes in x.

One of the simplest examples of a polynomial is a linear polynomial, for example:

$$P(x) = 3x + 2$$

This is a first-degree polynomial because the highest exponent of x is one. The graph of a linear polynomial is

6.1. INTRODUCTION TO POLYNOMIALS

a straight line, showing a constant rate of change of $P(x)$ with x.

Another example is a quadratic polynomial, such as:

$$P(x) = 4x^2 - 5x + 1$$

Quadratic polynomials have a degree of two, and their graphs are parabolas. They can have zero, one, or two real roots, depending on the discriminant ($b^2 - 4ac$ for a quadratic equation $ax^2 + bx + c$).

Understanding polynomials involves not just manipulating expressions but also grasping their fundamental properties. For instance, the degree of a polynomial can give us insights into the number of roots it might have, and its leading coefficient can help us understand its end behavior as x approaches infinity or negative infinity.

The operations on polynomials - addition, subtraction, multiplication, and division - follow set rules. When combining polynomials through these operations, the result is always another polynomial. This property makes them closed under these operations, which means that these operations do not produce outcomes outside the set of polynomials.

Polynomials are not just theoretical constructs in mathematics. They have practical applications in fields as diverse as physics, engineering, economics, and biology. Understanding how to work with polynomials, therefore, is not just an academic exercise but a skill that has real-world relevance. Whether calculating areas, modeling population growth, or designing complex algorithms, polynomials are a fundamental tool that professionals and researchers rely on.

In this chapter, we will explore the various types of polynomials, how to perform operations on them, and how they are used in solving equations and modeling real-world situations. Starting with the basics, we will gradually progress to more complex concepts, ensuring a deep understanding of polynomials and their applications.

6.2 Types of Polynomials

Polynomials, fundamentally, are expressions comprised of variables and coefficients, brought together through the operations of addition, subtraction, multiplication, and non-negative integer exponents of variables. Understanding the classification and hierarchy of polynomials is crucial for navigating the broader topics of algebra and serves as a foundation for the operations and applications discussed in subsequent sections. This segment elucidates the types of polynomials, focusing on their categorization based on the number of terms they contain and the degree of the polynomial.

The simplest variety of polynomial is the *monomial*, which contains only one term. A monomial can be a number, a variable, or a product of numbers and variables raised to a power. The general form of a monomial is ax^n, where a is the coefficient, n is the non-negative integer exponent, and x is the variable. Notably, the degree of a monomial is the exponent n that the variable is raised to. For instance, in $7x^3$, the coefficient is 7, and its degree is 3.

```
Example:
   Monomial: 4x^2
   Coefficient: 4
   Degree: 2
```

Following monomials, we encounter *binomials*, which, as the prefix bi- suggests, consist of two terms. A binomial

6.2. TYPES OF POLYNOMIALS

is the sum or difference of two monomials that are not like terms, meaning they do not have the same variable raised to the same power. An example of a binomial is $3x^2 + 4x$. Here, the degree of the binomial is the highest degree among its monomial terms, which, in this case, is 2.

```
Example:
    Binomial: 3x^2 + 4x
    Terms: 3x^2, 4x
    Degree: 2
```

Trinomials come next, comprising three terms. Analogous to binomials, the degree of a trinomial is determined by the highest degree of its terms. An example of a trinomial is $x^2 - 4x + 4$. Here, the terms are x^2, $-4x$, and 4, with the degree being 2.

```
Example:
    Trinomial: x^2 - 4x + 4
    Terms: x^2, -4x, 4
    Degree: 2
```

When polynomials contain more than three terms, they are generally referred to by the degree of the polynomial or simply as *polynomials*. The degree of a polynomial is pivotal because it directs many properties of the polynomial, including the behavior of its graph and the number of roots it can possess. The overall degree of a polynomial is identified by the term with the highest degree. For instance, the polynomial $2x^4 - 3x^3 + x - 9$ has a degree of 4, the highest exponent present in any term.

```
Example:
    Polynomial: 2x^4 - 3x^3 + x - 9
    Degree: 4
```

Furthermore, the classification of polynomials extends beyond the number of terms to include special types identified by their structure or the patterns they exhibit. Notable among these are the *homogeneous* and *cubic* polynomials. A homogeneous polynomial is characterized by all of its terms having the same total degree when considering the sum of the exponents of each term. Conversely, a cubic polynomial is defined by having a degree of 3, highlighting the importance of the polynomial's degree in its classification.

- Homogeneous polynomial example: $2x^3 + 3xy^2 + y^3$
- Cubic polynomial example: $x^3 - 6x^2 + 11x - 6$

In summary, the types of polynomials, classified based on the number of terms and the degree, serve as a foundation for understanding polynomial operations and their applications. This classification system aids in the systematic exploration of algebraic expressions and functions, paving the way for the advanced topics that follow in this chapter. Understanding these basic constructs enables students to navigate and manipulate polynomials with greater proficiency, a critical skill in the broader study of algebra.

6.3 Adding and Subtracting Polynomials

Adding and subtracting polynomials are foundational operations that allow us to combine or simplify polynomial expressions. Understanding these operations is crucial

6.3. ADDING AND SUBTRACTING POLYNOMIALS

for solving more complex algebraic expressions and equations involving polynomials. This section will provide a comprehensive guide on performing these operations, supported by detailed examples and clear explanations.

Polynomials are algebraic expressions that consist of variables, coefficients, and exponents. The basic form of a polynomial is $a_n x^n + a_{n-1} x^{n-1} + \cdots + a_1 x + a_0$, where a_n represents the coefficient for each term, and n is the degree of the polynomial, determined by the highest power of the variable x.

Adding Polynomials

To add polynomials, we align terms with similar degrees and then sum the coefficients of these like terms. This process is straightforward and follows the associative and commutative properties of addition.

- Step 1: Arrange both polynomials in descending order of their degrees.

- Step 2: Group like terms, which are terms that have the same variable raised to the same power.

- Step 3: Add the coefficients of like terms.

Example: Consider the polynomials $3x^2 + 2x + 1$ and $5x^2 - 3x + 4$. Adding these polynomials involves grouping like terms and summing their coefficients.

$$(3x^2 + 2x + 1)$$
$$+(5x^2 - 3x + 4)$$
$$=(3x^2 + 5x^2) + (2x - 3x) + (1 + 4)$$
$$=8x^2 - x + 5$$

```
Result: 8x^2 - x + 5
```

Subtracting Polynomials

Subtracting polynomials is similar to adding polynomials, with the primary difference being the introduction of subtraction. It's helpful to view subtraction as the addition of a negative.

- Step 1: Arrange both polynomials in descending order of their degrees.
- Step 2: Distribute the negative sign (if subtracting) to each term of the polynomial being subtracted.
- Step 3: Group like terms.
- Step 4: Subtract the coefficients of like terms.

Example: Consider subtracting the polynomial $2x^3 - x + 4$ from $3x^3 + 2x^2 - x - 1$.

$$(3x^3 + 2x^2 - x - 1)$$
$$-(2x^3 - x + 4)$$
$$=3x^3 - 2x^3 + 2x^2 - (-x) - 1 - 4$$
$$=x^3 + 2x^2 + x - 5$$

```
Result: x^3 + 2x^2 + x - 5
```

Practical Considerations

When adding or subtracting polynomials, it's essential to ensure that like terms are accurately identified and correctly grouped. A common mistake is to incorrectly group terms that are not similar, which can lead to errors in the calculation.

Additionally, understanding how to manipulate polynomials through addition and subtraction lays the groundwork for more complex operations such as multiplication and division, as well as solving equations involving polynomials. Mastery of these basic operations is crucial for success in higher-level algebra and calculus.

Exercises for Practice

1. Add $4x^3 + 3x^2 - x + 2$ to $-x^3 + 2x - 5$.
2. Subtract $2x^2 - 3x + 1$ from $5x^2 + x - 4$.
3. Add $x^4 - 3x^3 + x^2$ to $2x^3 - x^2 + 4x$.
4. Subtract $6x^4 - 4x^3 + 3x - 7$ from $8x^4 + x^3 - 5x^2 + x + 4$.

Solutions to these exercises offer additional practice in applying the principles of adding and subtracting polynomials, reinforcing understanding and skill in manipulating polynomial expressions.

6.4 Multiplying Polynomials

Multiplying polynomials is a fundamental operation that follows the distributive property of multiplication over addition. When multiplying two polynomials, each term of the first polynomial is multiplied by every term of the second polynomial. This process can be easily understood and applied through a variety of methods, including the vertical and horizontal methods, the FOIL method for binomials, and the area model.

Consider two polynomials $P(x) = a_n x^n + a_{n-1} x^{n-1} + \cdots + a_1 x + a_0$ and $Q(x) = b_m x^m + b_{m-1} x^{m-1} + \cdots + b_1 x + b_0$, where n and m are non-negative integers. The product of $P(x)$ and $Q(x)$ is obtained by applying the distributive property, resulting in a polynomial of degree $n + m$.

$$\begin{aligned} P(x) \cdot Q(x) = & (a_n x^n + a_{n-1} x^{n-1} + \cdots + a_1 x + a_0) \cdot \\ & (b_m x^m + b_{m-1} x^{m-1} + \cdots + b_1 x + b_0) \\ = & a_n b_m x^{n+m} + \cdots + (a_n b_0 + a_{n-1} b_1 + \cdots + a_0 b_m) \\ & + \cdots + a_0 b_0 \end{aligned}$$

The resulting coefficients are obtained by summing the products of the coefficients from $P(x)$ and $Q(x)$, corresponding to the exponents that add up to the same value.

Example: Multiply $(3x^2 + 2x + 1)$ by $(x + 4)$.

Applying the distributive property, we have:

$$\begin{aligned} (3x^2 + 2x + 1) \cdot (x + 4) = & \, 3x^2 \cdot x + 3x^2 \cdot 4 + 2x \cdot x + 2x \cdot 4 \\ & + 1 \cdot x + 1 \cdot 4 \\ = & \, 3x^3 + 12x^2 + 2x^2 + 8x + x + 4 \\ = & \, 3x^3 + 14x^2 + 9x + 4 \end{aligned}$$

This example illustrates the horizontal method of polynomial multiplication, where terms are aligned and multiplied accordingly.

6.4. MULTIPLYING POLYNOMIALS

An alternative approach is the FOIL method, which stands for First, Outer, Inner, Last, and is specifically used for multiplying binomials.

Example: Multiply $(x+3)(x+2)$.

Following the FOIL steps, we get:

$(x+3)(x+2) =$
$\underline{First}: x \cdot x = x^2$
$\underline{Outer}: x \cdot 2 = 2x$
$\underline{Inner}: 3 \cdot x = 3x$
$\underline{Last}: 3 \cdot 2 = 6$
Result: $x^2 + 2x + 3x + 6 = x^2 + 5x + 6$

The area model for multiplying polynomials involves drawing a rectangle, dividing it into smaller sections based on the number of terms in each polynomial, and filling in each section with the product of the corresponding terms. This visual representation can simplify the multiplication process, especially for polynomials with more than two terms.

$a_n x^n \cdot b_m x^m$	$\vdots$
$\vdots$	$a_0 \cdot b_0$

Multiplying polynomials involves utilizing the distributive property to combine like terms, resulting in a new polynomial with a degree that is the sum of the degrees of the polynomials being multiplied. By meticulously applying the appropriate method, whether it be the horizontal and vertical methods, the FOIL method, or the area model, students can accurately and efficiently multiply polynomials of varying degrees and complexity.

6.5 Special Products of Polynomials

Understanding the special products of polynomials is pivotal in simplifying and solving polynomial equations more efficiently. This section explores the formulae for the square of a binomial, the product of a sum and difference, and the cube of a binomial. These identities, once mastered, will considerably speed up polynomial multiplication without the need for lengthy multiplication processes.

Square of a Binomial: The square of a binomial follows the pattern $(a+b)^2 = a^2 + 2ab + b^2$ and $(a-b)^2 = a^2 - 2ab + b^2$. This means, when squaring a binomial, one squares the first term, multiplies the two terms together and doubles it for the middle term, and then squares the last term.

$$\text{For example, } (3x+4)^2 = (3x)^2 + 2(3x)(4) + 4^2$$
$$= 9x^2 + 24x + 16.$$

Product of Sum and Difference: The product of a sum and a difference follows the formula $(a+b)(a-b) = a^2 - b^2$. This property, also known as the difference of squares, shows that the product of a binomial and its conjugate results in the difference between the squares of the terms.

$$\text{For example, } (5x+3)(5x-3) = (5x)^2 - (3)^2$$
$$= 25x^2 - 9.$$

6.5. SPECIAL PRODUCTS OF POLYNOMIALS

Cube of a Binomial: The cube of a binomial is given by $(a+b)^3 = a^3+3a^2b+3ab^2+b^3$ and $(a-b)^3 = a^3-3a^2b+3ab^2-b^3$. When cubing a binomial, one cubes the first term, triples the product of the square of the first term and the second term, triples the product of the first term and the square of the second term, and finally, cubes the second term.

For example,
$$(2x-1)^3 = (2x)^3 - 3(2x)^2(1) + 3(2x)(1)^2 - 1^3$$
$$= 8x^3 - 12x^2 + 6x - 1.$$

Practice Problems: To solidify understanding, consider the following practice problems.

- Compute $(x+5)^2$.
- Determine the product of $(3a+2b)(3a-2b)$.
- Find $(2y-3)^3$.

Applying these special product formulas not only simplifies polynomial operations but also aids in recognizing patterns and relationships within polynomials, essential for further algebraic manipulations and problem-solving. Mastery of these identities will serve as a foundational tool in progressing through more complex polynomial functions and equations.

Students are encouraged to practice these identities with various polynomials to gain comfort and familiarity with their application. The efficacy of these special products becomes especially apparent in more advanced algebraic procedures, such as factoring polynomials and solving polynomial equations, where they can dramatically reduce computational effort.

6.6 Polynomial Long Division

Polynomial long division is a technique for dividing a polynomial by another polynomial of lower degree, similar in approach to the long division of numbers. This method is essential for simplifying expressions, solving polynomial equations, and analyzing polynomial functions. The process closely mirrors that of numerical long division but requires a keen understanding of algebraic manipulation.

To begin, let us consider the division of a polynomial $P(x)$ by another polynomial $d(x)$, where the degree of $P(x)$ is greater than or equal to the degree of $d(x)$. The objective is to find two polynomials, the quotient $q(x)$ and the remainder $r(x)$, such that

$$P(x) = d(x)q(x) + r(x),$$

with the degree of $r(x)$ less than the degree of $d(x)$.

The steps to perform polynomial long division are as follows:

- Arrange the terms of both the dividend $P(x)$ and the divisor $d(x)$ in descending order of their degrees. If any terms are missing, include them with a coefficient of 0.

- Divide the leading term of the dividend by the leading term of the divisor to obtain the first term of the quotient.

- Multiply the entire divisor by this first term of the quotient and write the result beneath the dividend, aligning like terms.

6.6. POLYNOMIAL LONG DIVISION

- Subtract this result from the dividend to find the remainder. If the degree of this remainder is greater than or equal to the degree of the divisor, repeat the process with the remainder as the new dividend.

- Continue this process until the degree of the remainder is less than the degree of the divisor. The quotient obtained through this process, along with the final remainder, represents the result of the division.

Let's illustrate this with an example. Consider dividing $P(x) = 2x^3 - 3x^2 + 4x - 5$ by $d(x) = x - 1$.

```
1. Arrange in descending order: Terms are already in order.
2. Divide the leading term of P(x) by the leading term of
   d(x): 2x^3 / x = 2x^2.
3. Multiply d(x) by 2x^2 and subtract from P(x):
     2x^3 - 3x^2 + 4x - 5
   - (2x^3 - 2x^2)
     ------------------
            -x^2 + 4x - 5
4. Repeat with the new dividend of -x^2 + 4x - 5:
   Divide -x^2 by x, obtaining -x. Multiply and subtract:
          -x^2 + 4x - 5
        - (-x^2 - x)
          ---------------
                  5x - 5
5. Finally, divide 5x by x, resulting in 5. Multiply and subtract:
          5x - 5
        - (5x - 5)
          ---------
                 0
```

The quotient is $q(x) = 2x^2 - x + 5$ and the remainder is $r(x) = 0$. Therefore,

$$2x^3 - 3x^2 + 4x - 5 = (x-1)(2x^2 - x + 5) + 0.$$

It is crucial to note that the remainder can sometimes be non-zero, indicating the division does not result in a polynomial but rather a polynomial plus a fraction. For instance, if dividing $2x^2 + 3x + 1$ by $x + 1$, the remainder is not zero, and the division represents an incomplete simplification.

Polynomial long division serves as a foundational tool in algebra for simplifying expressions and understanding the structure of polynomials. Its application extends to various fields, including calculus, where it aids in the integration of rational functions, and in number theory, as it mimics the division algorithm for integers. Mastery of polynomial long division equips students with the skills to navigate complex algebraic manipulations and understand the intricacies of polynomial functions.

6.7 Synthetic Division

Synthetic division is a simplified process for dividing a polynomial by a binomial of the form $x - c$. It reduces the cumbersome steps involved in polynomial long division to a more manageable algorithm, aiding in the efficiency and accuracy of computations. This method is particularly useful for dividing polynomials when the divisor is a first-degree polynomial.

To perform synthetic division, one must follow a series of steps, which can be thought of as an algorithm. The dividend's coefficients are written in a row, including zeros for any missing terms, which represents the polynomial's completeness in terms of its degree. The divisor must be in the form $x - c$, where c is a constant. If the divisor is not in this form, the problem must be reformulated so that it is.

- Bring down the leading coefficient to the row below at the start of the division.

6.7. SYNTHETIC DIVISION

- Multiply this number by c (the value from the divisor $x - c$) and write the result under the second coefficient.

- Add the numbers in this column.

- Repeat the multiplication and addition steps until all the coefficients have been processed.

An illustrative example can clarify the synthetic division process considerably:

Consider dividing $2x^3 - 6x^2 + 2x - 4$ by $x - 3$. Here, $c = 3$.

```
Step 1: Write down the coefficients: 2 -6 2 -4
Step 2: Bring down the 2.
Step 3: Multiply 2 by 3 (the c value) to get 6. Write this under the -6.
Step 4: Sum the column: -6 + 6 = 0. Write this down.
Step 5: Multiply 0 by 3 to get 0. Add this to 2 to get 2.
Step 6: Multiply 2 by 3 to get 6. Add this to -4 to get 2.
```

The final row now represents the quotient's coefficients, while the last number is the remainder. Therefore, the quotient is $2x^2 + 0x + 2$ or simply $2x^2 + 2$ with a remainder of 2. The division statement can be written as:

$$\frac{2x^3 - 6x^2 + 2x - 4}{x - 3} = 2x^2 + 2 + \frac{2}{x - 3}$$

One of the most significant advantages of synthetic division over long division is its compactness and speed, making it preferable for specific tasks such as finding polynomial roots or simplifying expressions before differentiation or integration in calculus.

While synthetic division offers a streamlined approach to division, it's crucial to recognize its limitations. It is applicable only when dividing by linear polynomials of the form $x - c$. For higher-degree divisors, polynomial long division remains necessary.

Synthetic division plays a crucial role when applying the Remainder Theorem and Factor Theorem, as it provides a quick method to evaluate polynomial functions at given points and determine factors of polynomials, respectively. Mastery of synthetic division thus allows for more efficient factorization, simplification, and solving of polynomial equations.

By integrating synthetic division into their repertoire, learners can enhance their polynomial manipulation skills, paving the way for deeper exploration into algebraic functions and their applications. It exemplifies the power of algebraic techniques in making seemingly complex procedures more accessible and manageable.

6.8 The Remainder Theorem and Factor Theorem

The Remainder Theorem and Factor Theorem are powerful tools in polynomial algebra, offering a deep understanding of the behaviors and properties of polynomial functions. These theorems are not only essential for theoretical mathematical pursuits but also have practical implications in solving polynomial equations and graphing polynomial functions.

The Remainder Theorem

The Remainder Theorem states that when a polynomial $f(x)$ is divided by a linear divisor of the form $(x - c)$, the

6.8. THE REMAINDER THEOREM AND FACTOR THEOREM

remainder of this division is equal to $f(c)$. This theorem provides a straightforward method for evaluating polynomials at specific points and is foundational for understanding the Factor Theorem.

Consider a polynomial $f(x)$ and a number c in the real number system. By the division algorithm for polynomials, we can express $f(x)$ as:

$$f(x) = (x - c)q(x) + r$$

where $q(x)$ is the quotient polynomial, and r is the remainder upon dividing by $(x - c)$. If the degree of $(x - c)$ is one, then the degree of r must be less than one, indicating that r is a constant.

To find the value of r, we substitute $x = c$ into the equation:

$$f(c) = (c - c)q(c) + r = 0 \cdot q(c) + r = r$$

Thus, when $f(x)$ is divided by $(x - c)$, the remainder is $f(c)$. This concept is illustrated in the following example.

Example:

Consider the polynomial $f(x) = x^3 - 4x^2 + 6x - 24$. Find the remainder when $f(x)$ is divided by $(x - 2)$.

Using the Remainder Theorem:

$$f(2) = 2^3 - 4(2)^2 + 6(2) - 24 = 8 - 16 + 12 - 24 = -20$$

Thus, the remainder is -20.

The Factor Theorem

Building on the Remainder Theorem, the Factor Theorem states that a polynomial $f(x)$ has a factor of $(x - c)$ if and only if $f(c) = 0$. In other words, $(x - c)$ is a factor of $f(x)$ if substituting $x = c$ into the polynomial yields zero. This theorem is invaluable for identifying the roots of a polynomial and understanding its factorization.

To see the Factor Theorem in action, let us consider any polynomial $f(x)$, which when divided by $(x - c)$ gives a remainder of zero. This implies that $f(x)$ can be written as:

$$f(x) = (x - c)q(x)$$

where $q(x)$ is some quotient polynomial. Since the remainder is zero, we know that $f(c) = 0$. Conversely, if $f(c) = 0$, it follows that $(x - c)$ must be a factor of $f(x)$.

Example:

Let $f(x) = x^2 - 5x + 6$. To determine whether $x - 2$ is a factor of $f(x)$, evaluate $f(2)$:

$$f(2) = 2^2 - 5(2) + 6 = 4 - 10 + 6 = 0$$

Since $f(2) = 0$, by the Factor Theorem, $x - 2$ is indeed a factor of $f(x)$.

Application of Theorems

These theorems are not just theoretical constructs but have direct applications in solving higher-degree polynomial equations, simplifying polynomial expressions, and understanding the graphical behavior of polynomial functions. For instance, if we can identify the roots of a polynomial through the Factor Theorem, we can then easily sketch the graph of the polynomial by marking its x-intercepts and analyzing the multiplicity of the roots.

Together, the Remainder and Factor Theorems are indispensable in the toolkit of anyone studying or working with polynomials. Through a combination of theory and worked examples, we have demonstrated how these theorems can be applied to solve problems effectively, offering both a deeper insight into the structure of polynomials and practical techniques for their analysis.

6.9 Solving Polynomial Equations

Solving polynomial equations is a fundamental aspect of algebra that involves finding the values of the variable for which the polynomial is equal to zero. These values are known as the roots or solutions of the equation. Polynomial equations can be of varying degrees, and the degree of the polynomial influences the methods that can be used for solving them as well as the number of solutions to expect.

Let us start by considering a polynomial equation of the form $P(x) = 0$, where $P(x)$ is a polynomial expression. The fundamental principle in solving polynomial equations is to express the polynomial in its factored form, if possible, and then apply the Zero Product Property. The Zero Product Property states that if the product of two or more expressions is zero, then at least one of the expressions must be zero.

Factoring Polynomials: The first step in solving polynomial equations is often to factor the polynomial, if feasible. This involves expressing the polynomial as a product of its factors. Common factoring techniques include factoring by grouping, using special product patterns, and the application of the Factor Theorem. Once a polynomial has been factored, the solutions can be found by setting each factor equal to zero and solving for the variable.

For example, consider the equation $(x - 3)(x + 5) = 0$.
Setting each factor equal to zero yields:
$x - 3 = 0$ or $x + 5 = 0$.
Solving for x gives the solutions: $x = 3$ and $x = -5$.

Using the Quadratic Formula: For quadratic polynomials (degree 2), if factoring is not straightforward, the Quadratic Formula can be used. Given a quadratic equation in the form $ax^2 + bx + c = 0$, the solutions can be found using the formula $x = \frac{-b \pm \sqrt{b^2 - 4ac}}{2a}$.

Example: Consider the equation $2x^2 - 4x - 6 = 0$. Applying the Quadratic Formula yields:

$$x = (4 \pm \sqrt{(-4)^2 - 4 * 2 * (-6)})/(2 * 2)$$
$$x = (4 \pm \sqrt{16 + 48})/4$$
$$x = (4 \pm \sqrt{64})/4$$
$$x = (4 \pm 8)/4$$

Thus, the solutions are x = 3 and x = -1.

Polynomial Division: For higher degree polynomials, Long Division or Synthetic Division may be used to simplify the polynomial before attempting to solve it. This

6.9. SOLVING POLYNOMIAL EQUATIONS

is particularly useful for identifying factors or applying the Remainder Theorem and Factor Theorem to find solutions.

The Rational Root Theorem: The Rational Root Theorem is a useful tool for identifying possible rational roots of polynomial equations, based on the relationship between the coefficients of the highest and lowest degree terms. It provides a set of potential rational solutions that can be tested to find the actual roots.

Graphical Methods: Graphing polynomial functions can provide a visual method for estimating the roots of a polynomial equation. The x-intercepts of the graph correspond to the solutions of the equation. This method, while not always precise, can offer valuable insights, especially when combined with other algebraic methods.

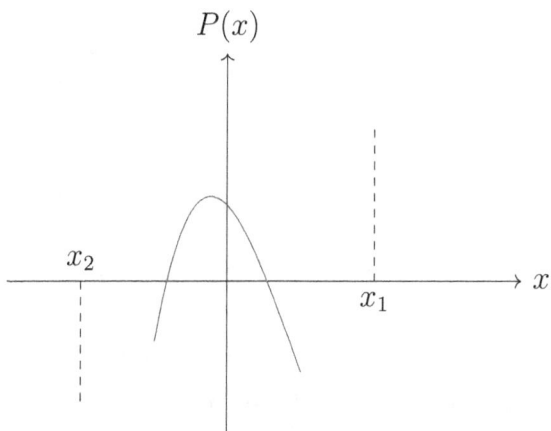

In solving polynomial equations, it is crucial to explore different methods and apply the most appropriate ones based on the degree of the polynomial and the form of the equation. Mastery of these concepts allows for the effective resolution of polynomial equations, paving the way for deeper exploration of algebraic functions and their applications.

6.10 Graphing Polynomial Functions

The process of graphing polynomial functions is a critical skill in algebra that enables the visualization of the behavior and properties of polynomials. This section is devoted to guiding students through the foundational steps necessary to graph polynomial functions effectively, emphasizing key characteristics such as end behavior, intercepts, and turning points that significantly impact the graph's shape.

Understanding the Basic Shape and End Behavior: Initially, it is crucial to recognize that the degree of the polynomial function influences its general shape and end behavior. For a polynomial function $f(x) = a_n x^n + a_{n-1} x^{n-1} + \cdots + a_1 x + a_0$, where $a_n \neq 0$ and n is a non-negative integer, the end behavior is determined by the leading term, $a_n x^n$.

$$\lim_{x \to +\infty} f(x) = \begin{cases} +\infty, & \text{if } a_n > 0 \text{ and } n \text{ is even,} \\ -\infty, & \text{if } a_n < 0 \text{ and } n \text{ is even,} \\ +\infty, & \text{if } a_n > 0 \text{ and } n \text{ is odd,} \\ -\infty, & \text{if } a_n < 0 \text{ and } n \text{ is odd.} \end{cases}$$

$$\lim_{x \to -\infty} f(x) = \begin{cases} +\infty, & \text{if } a_n > 0 \text{ and } n \text{ is even,} \\ -\infty, & \text{if } a_n < 0 \text{ and } n \text{ is even,} \\ -\infty, & \text{if } a_n > 0 \text{ and } n \text{ is odd,} \\ +\infty, & \text{if } a_n < 0 \text{ and } n \text{ is odd.} \end{cases}$$

Identifying Intercepts: Next, identifying x-intercepts and y-intercepts is essential to plot key points on the

6.10. GRAPHING POLYNOMIAL FUNCTIONS

graph. The x-intercepts are found by solving the equation $f(x) = 0$, and the y-intercept is located by evaluating $f(0)$.

Using Turning Points: A crucial concept in graphing polynomial functions is understanding turning points, which are points where the graph changes direction. The number of possible turning points in a polynomial function of degree n is at most $n - 1$.

Sketching the Graph: With an understanding of end behavior, intercepts, and turning points, a preliminary sketch of the graph can be constructed. Here, we integrate these concepts through a simple example.

Example: Consider the polynomial function $f(x) = x^3 - 4x^2 + 3x$.

Intercepts:

- To find the x-intercepts, solve $x^3 - 4x^2 + 3x = 0$, which simplifies to $x(x^2 - 4x + 3) = 0$. The solutions are $x = 0, 1, 3$.

- The y-intercept is found by evaluating $f(0) = 0$.

End Behavior: Since the leading term is x^3 and the leading coefficient is positive, the end behavior is such that as $x \to -\infty$, $f(x) \to -\infty$ and as $x \to +\infty$, $f(x) \to +\infty$.

Sketch: The intercepts and end behavior are plotted initially. Noting the maximum possible turning points hint at a potential graph shape, the actual plotting reveals the general trajectory of the function through these key points and behaviors.

```
Sketch of the graph (not shown here) would display the curve passing through
the intercepts (0, 0), (1, 0), and (3, 0), with ends heading towards negative
infinity on the left and positive infinity on the right.
```

Graphing polynomial functions involves a combination of analyzing the polynomial's degree, calculating intercepts, predicting end behavior, and pinpointing any turning points. Mastery of these steps allows for the creation of an accurate sketch of any polynomial function, providing invaluable insight into the function's properties and behavior.

6.11 Polynomial Inequalities

Understanding polynomial inequalities involves solving inequalities in which a polynomial expression is set to be greater than or less than zero. These inequalities share similarities with polynomial equations, but instead of finding exact solutions, the goal is to determine ranges of values that satisfy the inequality. Given the crucial role of polynomial inequalities in various areas of mathematics, including calculus and algebraic geometry, mastering them is essential. This section addresses the process of solving polynomial inequalities through a series of systematic steps, complemented by examples.

The general form of a polynomial inequality can be written as $P(x) > 0$, $P(x) \geq 0$, $P(x) < 0$, or $P(x) \leq 0$, where $P(x)$ is a polynomial in x. The approach to solving these inequalities involves several key steps: finding the roots of the associated polynomial equation, determining the sign of the polynomial in the intervals defined by these roots, and solving the inequality based on the sign of the polynomial in these intervals.

Step 1: *Find the roots of the associated polynomial equation.* This entails solving the equation $P(x) = 0$ to find the values of x where the polynomial equals zero. These roots

6.11. POLYNOMIAL INEQUALITIES

are crucial as they divide the real number line into intervals on which the sign of $P(x)$ can be examined.

```
Example: Solve P(x) = x^3 - 4x^2 + 3x = 0
Solution: x(x^2 - 4x + 3) = x(x - 1)(x - 3) = 0
Roots: x = 0, x = 1, x = 3
```

Step 2: *Determine the sign of the polynomial on the intervals defined by the roots.* This step involves selecting a test value within each interval created by the roots and substituting it into the polynomial to ascertain the sign of $P(x)$ in that interval.

```
Using the roots from the example above:
Choose test values -1, 0.5, 2, and 4 for the intervals (-∞, 0), (0, 1),
(1, 3), and (3, ∞), respectively.
P(-1) = 8   (Positive)
P(0.5) = -0.125 (Negative)
P(2) = 2    (Positive)
P(4) = 4    (Positive)
```

Step 3: *Solve the inequality.* Using the sign of the polynomial on the intervals, determine the intervals that satisfy the original inequality condition. For inequalities involving $>$ or $<$, exclude the roots, as the polynomial is not greater or less than zero at these points. For inequalities involving $\geq$ or $\leq$, include the roots.

```
For the inequality P(x) > 0, with P(x) = x^3 - 4x^2 + 3x:
The intervals where P(x) is positive are (-∞, 0), (1, 3), and (3, ∞).
```

It is crucial to provide a diagram to visually represent the steps and exhibit the intervals where the polynomial satisfies the initial inequality condition.

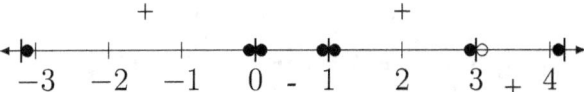

This visual representation, along with the steps outlined above, offers a structured approach to tackling polynomial inequalities, enabling students to identify solution intervals with greater ease. Mastery of these techniques enhances problem-solving skills in algebra and beyond, highlighting the importance of understanding and applying polynomial inequalities.

6.12 Applications of Polynomials

Polynomials are not merely mathematical expressions of interest within the realm of abstract algebra; they find applications in various disciplines including physics, engineering, economics, and even computer science. This section aims to elucidate some of these applications, providing a tangible connection between the theoretical constructs of polynomials and their utility in solving real-world problems.

One of the most common applications of polynomials is in physics, particularly in the study of projectile motion. The trajectory of an object in projectile motion can be modeled by a quadratic polynomial. Consider an object thrown into the air; its height h at any time t can be represented by the equation $h(t) = -gt^2 + vt + h_0$, where g is the acceleration due to gravity, v is the initial velocity, and h_0 is the initial height. This equation is a polynomial of degree 2, also known as a quadratic polynomial.

```
Example 1: If an object is thrown upwards with an initial velocity of
20 m/s from a height of 5 meters, its height after t seconds can be
modeled by the equation h(t) = -9.8t^2 + 20t + 5.
```

In engineering, polynomials are used in control system design and analysis. The stability of a system can be de-

6.12. APPLICATIONS OF POLYNOMIALS

termined by examining the roots of its characteristic polynomial. A system is considered stable if all the roots of its characteristic polynomial have negative real parts.

Economics provides another fertile ground for the application of polynomials. Demand curves, which illustrate the relationship between the price of a commodity and the quantity demanded, can often be modeled using polynomial functions. Similarly, cost functions, expressing the total cost of production as a function of the quantity produced, frequently employ polynomial expressions.

```
Example 2: A company's total cost \(C\) in dollars for producing \(x\)
units of a product can be modeled by the polynomial
  C(x) = 0.01x^3 - 0.6x^2 + 30x + 5000.
```

Computer science applications of polynomials include algorithms for error detection and correction in data transmission, known as cyclic redundancy checks (CRC). These algorithms utilize polynomial division to detect changes to raw data, thereby ensuring the integrity of the data being transmitted or stored.

Polynomials also play a crucial role in cryptography, particularly in the construction of public key cryptosystems such as RSA. In such systems, the security relies on the difficulty of factoring large polynomials into their prime factors.

Moreover, in modeling natural phenomena, polynomials are utilized to approximate complex functions. For example, the population growth of a species can be modeled using a polynomial function, where the rate of growth depends on the current population size in a non-linear manner.

```
Example 3: The population \(P\) of a certain species over time \(t\)
can be modeled by the polynomial P(t) = 0.3t^4 - 2t^3 + 10t^2 + 100.
```

In numerical methods, polynomials are used in interpolation techniques such as Lagrange and Newton interpolation, which are essential for estimating values of a function based on its known values at specific points.

Lastly, in optimization problems, polynomials are used to determine the maximum or minimum values of functions. This is particularly prevalent in operations research and management science, where cost minimization or profit maximization problems are modeled using polynomials.

The applications of polynomials extend far beyond the confines of mathematics textbooks. They are indispensable tools in various scientific disciplines, offering solutions to an array of complex problems. Through the understanding and manipulation of polynomials, one can gain insights into the mechanisms governing natural and man-made phenomena, thereby highlighting the profound impact of mathematical theory on practical, real-world applications.

Chapter 7

Factoring Polynomials

This chapter centers on factoring polynomials, a fundamental skill in algebra that simplifies the process of solving polynomial equations. The techniques discussed include factoring out the greatest common factor, factoring by grouping, special factoring formulas, and factoring trinomials. It also introduces advanced strategies for factoring such as the difference of squares, sum and difference of cubes, and solving equations by factoring. Through careful explanation and numerous examples, students will become proficient at recognizing and applying the appropriate factoring technique, paving the way for solving complex polynomial equations and deepening their understanding of algebraic structures.

7.1 Introduction to Factoring Polynomials

Factoring polynomials is a crucial concept in algebra that involves breaking down a polynomial into a product of simpler polynomials. This foundational skill is not only essential for solving polynomial equations but also serves as a cornerstone for further studies in algebra and calculus. The purpose of this section is to introduce the concept of factoring polynomials, elucidate its importance, and outline the basic strategies that will be explored in detail in subsequent sections of this chapter.

At its core, the process of factoring transforms a complex polynomial expression into a product of simpler factors. This simplification can make it easier to understand the properties of the original polynomial, including its roots, and can be instrumental in solving equations and inequalities.

The simplest form of factoring involves identifying and factoring out the greatest common factor (GCF) among the terms of the polynomial. This approach is based on the distributive law of multiplication over addition and is the first strategy that one should apply when attempting to factor any polynomial.

Consider the polynomial $8x^3 + 4x^2 - 16x$. The greatest common factor of these terms is $4x$, as each term is divisible by $4x$. Factoring out $4x$, we get:

$$8x^3 + 4x^2 - 16x = 4x(2x^2 + x - 4)$$

Following the extraction of the GCF, the polynomial inside the brackets may sometimes be further factored us-

7.1. INTRODUCTION TO FACTORING POLYNOMIALS

ing various techniques, as we will see later in this chapter.

Another fundamental approach to factoring polynomials is "factoring by grouping," which involves rearranging and grouping terms in a way that allows us to apply the distributive law. This method is particularly useful for polynomials with four or more terms.

For instance, consider the polynomial $x^3 + 3x^2 + 2x + 6$. By grouping the terms in pairs and factoring out the common factor in each, we achieve:

$$\begin{aligned} x^3 + 3x^2 + 2x + 6 &= (x^3 + 3x^2) + (2x + 6) \\ &= x^2(x + 3) + 2(x + 3) \\ &= (x^2 + 2)(x + 3) \end{aligned}$$

The process of factoring becomes more nuanced and challenging when dealing with trinomials and higher-degree polynomials. Methodologies for tackling these situations, including the use of special factoring formulas and strategies for factoring trinomials, will be discussed later in this chapter.

Significantly, mastering the art of factoring polynomials is not merely an academic exercise; it has practical applications in various fields of science, engineering, and finance. For example, factoring polynomial functions can reveal intercepts and critical points, providing crucial insights into the behavior of physical systems described by those functions.

In sum, the ability to factor polynomials effectively opens up new insights into the structure of algebraic expressions and prepares the ground for advancing further into the realm of mathematical analysis. The strategies covered in this chapter lay a solid foundation upon which complex problem-solving skills can be built, enabling students to navigate the challenges of algebra with confidence.

To facilitate a comprehensive understanding, each section that follows will delve into specific factoring techniques, supported by carefully selected examples and exercises designed to reinforce the concepts discussed. Through a methodical approach, students will gain proficiency in recognizing the most appropriate factoring technique for a given polynomial, significantly simplifying the process of solving polynomial equations.

7.2 Factoring out the Greatest Common Factor (GCF)

The process of factoring out the Greatest Common Factor (GCF) from a polynomial is a foundational technique in algebra. It involves identifying the highest common factor shared by all the terms in the polynomial and then dividing each term by this factor. The result simplifies the original polynomial into a product of the GCF and another polynomial.

Definition: The GCF of a polynomial is the highest degree of a monomial that divides each term of the polynomial without leaving a remainder.

To elucidate the concept, consider a polynomial $ax^n + bx^m$. The GCF of this polynomial is cx^k, where c is the greatest common factor of a and b, and k is the lesser of n and m.

Procedure for Factoring out the GCF:

- Identify the GCF of the coefficients of all terms in

7.2. FACTORING OUT THE GREATEST COMMON FACTOR (GCF)

the polynomial.

- Determine the smallest power of the variable that appears in all terms.
- Express the GCF as the product of these factors.
- Divide each term of the polynomial by the GCF.
- Rewrite the polynomial as the product of the GCF and the quotient obtained in the previous step.

Example 1: Factor $6x^3 + 9x^2$.

The first step is to identify the GCF of the coefficients 6 and 9, which is 3. Next, observe the variables. The smallest power of x in both terms is x^2. Therefore, the GCF is $3x^2$.

Divide each term by $3x^2$:

$$6x^3 = 3x^2 \cdot 2x,$$
$$9x^2 = 3x^2 \cdot 3.$$

The factored form of the polynomial is:

```
3x^2(2x + 3)
```

Example 2: Factor $12a^5b^3 + 18a^3b^4 - 24a^2b^5$.

The coefficients 12, 18, and -24 have a GCF of 6. The smallest power of a in all terms is a^2 and the smallest power of b is b^3. Hence, the GCF is $6a^2b^3$.

Dividing each term by $6a^2b^3$ yields:

$$12a^5b^3 = 6a^2b^3 \cdot 2a^3,$$
$$18a^3b^4 = 6a^2b^3 \cdot 3b,$$
$$-24a^2b^5 = 6a^2b^3 \cdot (-4b^2).$$

Therefore, the factored form is:

$$6a^2b^3(2a^3 + 3b - 4b^2)$$

Importance of Factoring out the GCF:

Factoring out the GCF simplifies polynomials, making them easier to manipulate in further operations such as division, further factoring, or solving equations. It is often the first step in many algebraic procedures because it reduces the degree and complexity of the polynomial, making the next steps more manageable.

Moreover, factoring out the GCF is critical in solving polynomial equations. When a polynomial is set equal to zero, factoring it completely can reveal its roots directly. This process begins effectively with extracting the GCF.

In summary, mastering the technique of factoring out the GCF is not only essential for simplifying and solving polynomial equations but also forms the basis for understanding more advanced algebraic concepts. Practice with a variety of polynomials is advised to become proficient in identifying and extracting the GCF swiftly and accurately.

7.3 Factoring by Grouping

Factoring by grouping is a technique used for factoring polynomials that are not easily factorable using other simpler methods such as factoring out the Greatest Common Factor (GCF) or applying special factoring formulas. The method is particularly useful for polynomials of four terms, although it can be adapted for polynomials with a different number of terms as long as they can be arranged into groups where factoring can be applied. The essence of factoring by grouping is to divide the polynomial into groups, usually of two terms each, factor out the GCF from each group, and then look for a common factor that can be factored out from the resulting expression.

The process of factoring by grouping involves the following steps:

- Arrange the terms of the polynomial so that those with common factors are next to each other.
- Divide the polynomial into groups, ideally of two terms each.
- Factor out the GCF from each group.
- If a common binomial factor emerges, factor it out.

To solidify understanding, consider the polynomial $8x^3 + 4x^2 - 2x - 1$. The goal is to factor this polynomial by grouping. Follow the steps outlined:

1. The polynomial can naturally be arranged into two groups of two terms each: $8x^3 + 4x^2$ and $-2x - 1$.

2. Factor out the GCF from each group. For the first group, the GCF is $4x^2$, and for the second group, it is -1. After factoring, the polynomial can be written as $4x^2(2x+1) - 1(2x+1)$.

3. Notice that $2x+1$ is a common factor between the two groups. Factoring this common binomial out results in $4x^2(2x+1) - 1(2x+1) = (2x+1)(4x^2-1)$.

4. The resulting expression contains a difference of squares, $4x^2 - 1$, which can be further factored as $(2x)^2 - (1)^2 = (2x+1)(2x-1)$. Therefore, the fully factored form of the original polynomial is $(2x+1)(2x+1)(2x-1)$.

Fully factored form: $(2x+1)^2(2x-1)$

This example illustrates the effectiveness of factoring by grouping as an intermediate step in simplifying complex polynomials. The technique reveals factors that are not immediately apparent, showing its utility in polynomial division and equation-solving processes.

Factoring by grouping is, however, conditionally applicable. It demands that the polynomial be rearrangeable into groups where a common factor exists or emerges after the GCF is factored out from each group. This condition emphasizes the importance of exploring various groupings and arrangements of terms, as the initial grouping might not immediately reveal a common factor. Persistence and practice are vital in mastering this technique, as is sensitivity to the structure of polynomials.

In teaching and applying factoring by grouping, emphasize flexibility in strategy and adaptability in approach. Encourage students to trial different groupings and not to hesitate in utilizing auxiliary steps, such as further factoring of resultant expressions, to achieve a fully factored form. Success in factoring by grouping significantly enhances algebraic manipulation skills, deepening understanding of polynomials and their properties.

7.4 Factoring Trinomials

Factoring trinomials is a critical component of algebra that involves rewriting a trinomial as the product of two or more polynomials. A trinomial is a polynomial with three terms, and in many cases of interest, it is quadratic, having the general form $ax^2 + bx + c$, where a, b, and c are constants. This section delves into techniques for factoring trinomials, focusing first on cases where $a = 1$, followed by more complicated scenarios where $a \neq 1$.

The essence of factoring lies in the search for two binomials $(dx+e)(fx+g)$ whose product is the original trinomial. Success hinges on the ability to discern the multiplicands through insight, practice, and systematic methods.

Factoring Trinomials with $a = 1$:

Let us commence with trinomials where the leading coefficient $a = 1$. This implies our trinomial takes the form $x^2 + bx + c$. The goal is to decompose it into two binomials $(x + d)(x + e)$.

To achieve this, we must find two integers d and e satisfying two conditions:

- The sum of d and e must be equal to b, the coefficient of the linear term.

- The product of d and e must be equal to c, the constant term.

Example 1: Consider the trinomial $x^2 + 5x + 6$. Here, we seek d and e such that $d + e = 5$ and $de = 6$. Through inspection or trial and error, we identify $d = 3$ and $e = 2$. Thus, $x^2 + 5x + 6 = (x+3)(x+2)$.

Factoring Trinomials with $a \neq 1$:

When dealing with trinomials of the form $ax^2 + bx + c$ where $a \neq 1$, the complexity increases. A systematic approach known as the "ac method" can be employed to navigate this challenge.

The "ac method" involves three steps:

- Multiply a and c to form a product termed ac.

- Search for two numbers that sum up to b (the coefficient of x) and multiply to ac.

- Use these numbers to express bx as the sum of two terms, which can then be factored by grouping.

Example 2: Consider $6x^2 + 11x + 3$. In this case, $a = 6$, $b = 11$, and $c = 3$. Multiplying a and c yields $ac = 18$. We seek two numbers whose product is 18 and whose sum is 11. These numbers are 9 and 2. Thus, $11x = 9x + 2x$ and the trinomial can be written as $6x^2 + 9x + 2x + 3$. Factoring by grouping, we get $3x(2x+3) + (2x+3)$, leading to $(2x+3)(3x+1)$.

Special Situations:

In some instances, trinomials can be factored into binomials that involve subtraction or complex coefficients. Identifying these patterns requires practice and a solid understanding of number properties.

For more challenging trinomials, alternative strategies such as completing the square or using the quadratic formula may prove more efficient. These methods also underscore the interconnections between different algebraic techniques and foster a deeper comprehension of polynomial structures.

Mastering the art of factoring trinomials is a cornerstone of algebra. It enables students to simplify expressions, solve quadratic equations, and understand the behavior of polynomial functions. The steps illustrated here, from handling simple cases with $a = 1$ to maneuvering through more complex instances where $a \neq 1$, form a robust foundation. Armed with these techniques, learners can approach factoring with confidence, aware that persistence and practice unlock the elegance and utility of algebra.

The next section will segue into special factoring formulas which further expand our factoring repertoire, tackling unique polynomial configurations with tailored strategies.

7.5 Special Factoring Formulas

Understanding special factoring formulas is crucial for efficiently simplifying and solving polynomial equations. This section explores three key formulas: the difference of squares, the sum and difference of cubes, and the square of a binomial. Mastery of these formulas will enable students to factor certain polynomials almost instantly, a skill that is invaluable in both algebra and calculus.

The difference of squares formula factors expressions of the form $a^2 - b^2$. It is given by:

$$a^2 - b^2 = (a+b)(a-b)$$

This formula is applicable whenever a polynomial is a difference (subtraction) between two perfect squares. To apply this formula, it is essential to recognize when a term is a perfect square and to correctly identify a and b.

Example: Factor $x^2 - 9$.

```
x^2 - 9 = (x)^2 - (3)^2 = (x + 3)(x - 3)
```

Next, we discuss the sum and difference of cubes. These formulas factor expressions of the form $a^3 + b^3$ and $a^3 - b^3$, which are not as immediately obvious as the difference of squares. The formulas are as follows:

$$a^3 + b^3 = (a+b)(a^2 - ab + b^2)$$
$$a^3 - b^3 = (a-b)(a^2 + ab + b^2)$$

These factorizations involve recognizing the cubic terms and applying the respective formula. The resulting factors include a binomial term and a trinomial term.

Example: Factor $x^3 - 27$.

```
x^3 - 27 = x^3 - (3)^3 = (x - 3)(x^2 + 3x + 9)
```

The final special formula considered here is the square of a binomial. It factors expressions of the form $(a+b)^2$ or $(a-b)^2$ directly without expansion. The formulas are given by:

$$(a+b)^2 = a^2 + 2ab + b^2$$
$$(a-b)^2 = a^2 - 2ab + b^2$$

7.5. SPECIAL FACTORING FORMULAS

This factoring approach is particularly useful in reverse when expanding binomials. Recognizing the specific pattern of coefficients can help in rewriting squared binomials without multiplication.

Example: Rewrite $(x+5)^2$ without expanding.

$$(x + 5)^2 = x^2 + 2(x)(5) + 5^2 = x^2 + 10x + 25$$

In practice, these special factoring formulas not only save time but also enhance the understanding of polynomial structures. Recognizing patterns in polynomials is a key skill in algebra, and these formulas are foundational in developing that skill. By applying these formulas, students can factor more complex polynomials, solve polynomial equations more efficiently, and better understand the relationship between algebraic expressions.

To solidify understanding, it is beneficial to solve practice problems that involve recognizing which special factoring formula to use and applying it correctly. Moreover, combining these formulas with other factoring techniques previously discussed can tackle a broad range of polynomial equations, demonstrating the versatile nature of algebra.

In summary, the special factoring formulas serve as powerful tools for simplifying and solving polynomials. Their consistent application reinforces the interconnectedness of algebraic concepts and fosters a deeper comprehension of mathematical reasoning and problem-solving strategies.

7.6 Factoring the Difference of Squares

The principle of factoring the difference of squares is a prevalent technique in algebra that allows us to factor polynomials of the form $a^2 - b^2$. This method leverages the identity $a^2 - b^2 = (a-b)(a+b)$, which states that the difference of two squares is equal to the product of the sum and difference of the two square roots.

To apply this technique effectively, it is crucial first to recognize a polynomial that fits the pattern of a difference of squares. Specifically, we are looking for an expression where both terms are perfect squares separated by a subtraction sign. Once identified, we can proceed to apply the identity to factor the polynomial.

$$a^2 - b^2 = (a-b)(a+b)$$

For example, consider the polynomial $x^2 - 9$. This expression can be seen as the difference of squares since both x^2 and 9 are perfect squares (x^2 is the square of x, and 9 is the square of 3). Applying the difference of squares formula gives us:

$$\begin{aligned} x^2 - 9 &= x^2 - 3^2 \\ &= (x-3)(x+3) \end{aligned}$$

Here, we substituted a with x and b with 3 to apply the identity directly. The factored form $(x-3)(x+3)$ can then be used for further algebraic manipulations or to find the roots of the equation $x^2 - 9 = 0$.

The difference of squares formula is not only applicable to monomial terms but can also be applied to polynomial

7.6. FACTORING THE DIFFERENCE OF SQUARES

terms that are perfect squares. For instance, consider the expression $(2x^2 + 6x)^2 - (4x)^2$. Both terms are perfect squares, and the expression is in the form $a^2 - b^2$, where $a = 2x^2 + 6x$ and $b = 4x$. Utilizing the factoring formula, we get:

$$(2x^2 + 6x)^2 - (4x)^2 = ((2x^2 + 6x) - 4x)((2x^2 + 6x) + 4x)$$
$$= (2x^2 + 2x)(2x^2 + 10x)$$

This shows how the difference of squares factoring can be extended to expressions involving polynomials of higher degrees, as long as the terms adhere to the perfect square criteria.

One of the compelling features of factoring the difference of squares is its applicability in simplifying expressions and solving equations. For example, when solving the equation $x^2 - 16 = 0$, recognizing that 16 is a perfect square allows us to factor the equation as $(x-4)(x+4) = 0$, which leads to the solutions $x = 4$ and $x = -4$.

In addition to solving equations, the difference of squares factoring is crucial in simplifying algebraic fractions. For instance, the expression $\frac{x^2-16}{x+4}$ can be simplified by factoring the numerator as $(x-4)(x+4)$ and then canceling out the $(x + 4)$ term in the numerator with the denominator, resulting in $\frac{x-4}{1}$ or simply $x - 4$.

```
Final example expression: x - 4
```

To summarize, mastering the difference of squares technique is fundamental in algebra. It enables the factorization of specific polynomials, simplifies expressions, and aids in solving equations. Recognizing expressions that fit the criteria for difference of squares and applying the identity $a^2 - b^2 = (a-b)(a+b)$ can streamline many algebraic processes.

7.7 Factoring the Sum and Difference of Cubes

Factoring the sum and difference of cubes is a crucial technique in simplifying complex algebraic expressions and solving polynomial equations. Unlike the more straightforward methods of factoring quadratics or applying the distributive property, the sum and difference of cubes require recognizing specific patterns and applying dedicated formulas.

To understand the essence of these formulas, let's start by stating them explicitly. For any two variables a and b, the sum of cubes is factored as follows:

$$a^3 + b^3 = (a+b)(a^2 - ab + b^2)$$

Similarly, the difference of cubes is factored using the formula:

$$a^3 - b^3 = (a-b)(a^2 + ab + b^2)$$

These formulas might initially seem abstract, but their application becomes more apparent through demonstration with examples.

Example 1: Factor $x^3 + 27$.

```
Step 1: Identify a and b:
Here, a = x and b = 3, since 27 is $3^3$.

Step 2: Apply the sum of cubes formula:
Thus, $x^3 + 27 = (x + 3)(x^2 - 3x + 9)$.
```

7.7. FACTORING THE SUM AND DIFFERENCE OF CUBES

Example 2: Factor $8y^3 - 64$.

Step 1: Identify a and b:

Here, a = 2y and b = 4, since $8y^3$ is $(2y)^3$ and 64 is 4^3.

Step 2: Apply the difference of cubes formula:

Thus, $8y^3 - 64 = (2y - 4)(4y^2 + 8y + 16)$.

When applying these formulas, it is essential to recognize the structure of the cubic terms. This recognition often involves seeing beyond the initial expression to identify perfect cubes and how they can be represented as a^3 or b^3.

Beyond simplifying expressions, factoring by sum and difference of cubes is strategically important in solving polynomial equations where the polynomial is in cubic form. Once the polynomial is factored, applying the Zero Product Property often allows for finding the roots of the equation.

Example 3: Solve $x^3 - 8 = 0$ **for** x.

Step 1: Factor the equation using the difference of cubes formula: $x^3 - 8 = (x - 2)(x^2 + 2x + 4)$.

Step 2: Set each factor equal to zero:

- For the first factor: $x - 2 = 0$, which implies $x = 2$.

- For the second factor: $x^2 + 2x + 4 = 0$, we find that it does not factor nicely, and its roots are not real numbers.

Therefore, the only real solution to the equation is $x = 2$.

A key insight into mastering the factoring of sum and difference of cubes is the practice with a diversity of expressions. This not only sharpens the skill of recognizing cubic forms but also deepens understanding of the structure and behavior of polynomial equations.

Practice Problem: Factor the following expressions:

- $x^3 + 125$
- $27x^3 - 1$
- $64y^3 + 8$

Factoring the sum and difference of cubes extends our toolkit for manipulating and solving polynomial equations. Its utility is most pronounced in its ability to break down complex cubic polynomials into simpler, more manageable forms, thereby facilitating a deeper comprehension of algebraic operations and the underpinning mathematical relationships.

7.8 Solving Polynomial Equations by Factoring

Solving polynomial equations by factoring is a pivotal technique in algebra, enabling the transformation of complex equations into simpler, solvable forms. This method leverages the Zero Product Property, which stipulates that if the product of two or more factors is zero, then at least one of the factors must be zero. This property is the cornerstone of solving polynomial equations through factoring.

To solve polynomial equations by factoring, one must first ensure the equation is set to zero. This establishes a foundation for applying the Zero Product Property effectively. Following this, the next step involves factoring the polynomial completely.

$$ax^2 + bx + c = 0$$

7.8. SOLVING POLYNOMIAL EQUATIONS BY FACTORING

This quadratic equation is a basic form where a, b, and c are constants. The process of factoring converts this equation into the form $(px+q)(rx+s) = 0$, where p, q, r, and s are also constants obtained through the factoring process.

Example 1: Consider the quadratic equation $x^2 - 5x + 6 = 0$.

To factor this equation, we seek two numbers whose product is equal to 6 (the constant term) and whose sum is -5 (the coefficient of the middle term). These numbers are -2 and -3. Thus, the factored form is:

$$(x - 2)(x - 3) = 0$$

Applying the Zero Product Property, we set each factor equal to zero:

x - 2 = 0   or   x - 3 = 0

Solving for x in each case gives:

x = 2   or   x = 3

This demonstrates that the solutions to the original equation are $x = 2$ and $x = 3$.

Advanced Factoring Techniques: For higher-degree polynomials, such as cubic equations ($ax^3 + bx^2 + cx + d = 0$), more sophisticated factoring techniques may be required, including the Rational Root Theorem and synthetic division. These methods assist in breaking down the polynomial into factors that can be solved using the Zero Product Property.

In solving real-world problems, factoring polynomials offers a method to find unknown quantities when the relationships between quantities are polynomial equations. For instance, in physics, factoring can be used to solve for time or distance when equations involve quadratic expressions related to acceleration, initial velocity, and initial position.

The importance of practice cannot be overstated in mastering the technique of solving polynomial equations by factoring. Varied exercises, ranging in complexity, provide the necessary experience to recognize the most efficient factoring method promptly. Moreover, applying these skills in practical scenarios enhances understanding and retention, enabling students to approach polynomial equations with confidence.

To summarize, solving polynomial equations by factoring is an essential algebraic skill that simplifies complex equations into a solvable format. By setting the polynomial equation to zero, factoring completely, and applying the Zero Product Property, one can find the solutions to the equation. Mastery of this technique is fundamental in progressing in algebra, offering a foundational tool for mathematical problem-solving across various applications.

7.9 Applications of Factoring Polynomials

Factoring polynomials is more than just a procedural skill; it has significant implications in various mathematical endeavours and real-world applications. This section explores the practicality of factoring in fields such as en-

7.9. APPLICATIONS OF FACTORING POLYNOMIALS

gineering, physics, and optimization problems. Understanding these applications highlights the relevance of algebra in solving tangible issues.

Optimization Problems: One of the most common applications of factoring polynomials is in optimization, where one seeks to determine the maximum or minimum values of a function. For instance, in manufacturing, a company might want to know the dimensions of a package that maximize volume while minimizing the cost of materials. Through the creation of a polynomial that represents the volume or cost, factoring can reveal the critical points where these values are optimized.

Consider the volume V of a box with a square base of side length x and height h, given by the polynomial $V(x) = x^2 h$. If the material for the sides costs more per unit area than the base and top, one can construct a cost function and use factoring to minimize cost for a given volume.

Physics and Engineering: In physics, factoring polynomials can be vital in kinematics, where the motion of objects is described algebraically. The position of an object, given as a function of time, can often be represented by a polynomial equation. Factoring these equations allows for the determination of specific times when an object will reach certain positions.

Similarly, in engineering, structural integrity problems might be expressed in terms of polynomial equations. Factoring these equations can help in identifying stress points and potential failure modes, aiding in the design of safer and more efficient structures.

Solving Differential Equations: Differential equations are fundamental in engineering and physics to describe how quantities change over time. Many techniques for solving differential equations involve factoring characteristic polynomials derived from the differential equations. By factoring these polynomials, solutions to the differential equations can be obtained, which describe the behavior of physical systems.

Cryptography: A surprising application of polynomial factoring is found in cryptography, particularly in algorithms like RSA, which rely on the difficulty of factoring large polynomials to secure data. In this context, factoring enables the encryption and decryption processes by leveraging the properties of prime numbers.

Root Finding and Graph Analysis: Factoring polynomials is crucial in finding the roots of equations, which correspond to the x-intercepts of the graph of the polynomial function. This information is invaluable in multiple disciplines, including economics to find break-even points, in chemistry to understand reaction rates, and in physics to analyze motion.

For example, consider a polynomial $P(x) = 2x^3 - 9x^2 + 7x - 1$. Factoring $P(x)$ allows us to find its roots, which could represent critical points in an application like maximizing profit in a business model.

Through these examples, it is clear that the ability to factor polynomials is not just an academic exercise but a powerful tool in both theoretical and practical problem solving. From optimizing processes in manufacturing to decrypting secure communications, the applications of factoring span a broad spectrum of disciplines. Mastery of this skill set provides a foundation not only for advanced mathematical studies but also for a wide array of professional and technical careers where problem-solving is paramount.

7.10 The Rational Root Theorem

One of the foundational tools in factoring polynomials, especially in finding the roots of polynomial equations, is the Rational Root Theorem. This theorem offers a systematic way to list all possible rational roots of a polynomial equation, thereby simplifying the search for actual roots. It is particularly useful in cases where the polynomials are not easily factorable using other techniques discussed in previous sections.

The statement of the Rational Root Theorem is as follows: For a polynomial equation $p(x) = a_n x^n + a_{n-1} x^{n-1} + \ldots + a_1 x + a_0 = 0$, where all coefficients a_i are integers and the leading coefficient a_n and constant term a_0 are non-zero, any rational root of the form $\frac{p}{q}$, where p and q are coprime integers (they have no common factors other than ± 1), must satisfy the following conditions:

- p is a factor of the constant term a_0,
- q is a factor of the leading coefficient a_n.

Based on these conditions, we can generate a list of all possible rational roots of the polynomial equation by taking all combinations of factors of a_0 and a_n, including both positive and negative values. This list can serve as a guide to systematically test potential roots through direct substitution into the polynomial equation or through synthetic division.

Let us demonstrate the practical application of the Rational Root Theorem with an example:

Consider the polynomial equation $2x^3 - 3x^2 - 5x + 6 = 0$. Here, the leading coefficient $a_n = 2$ and the constant term $a_0 = 6$. We first list the factors of these:

```
Factors of 6 (a_0): ±1, ±2, ±3, ±6
Factors of 2 (a_n): ±1, ±2
```

Based on the theorem, the possible rational roots $\frac{p}{q}$ are all combinations of these factors, specifically:

```
Possible rational roots: ±1/1, ±2/1, ±3/1, ±6/1, ±1/2, ±2/2, ±3/2, ±6/2
Simplified, this results in: ±1, ±2, ±3, ±6, ±1/2, ±3/2
```

To find the actual roots, each of these candidates can be tested either by plugging into the polynomial equation or using synthetic division. After testing, if the result is zero, the candidate is a root.

The Rational Root Theorem is vital not only because it helps in identifying potential roots but also because it highlights the interplay between the coefficients of a polynomial and its roots. This interplay is central to understanding more complex algebraic structures and techniques for factoring polynomials.

In addition to solving polynomial equations, the Rational Root Theorem can also play a role in constructing polynomials given a set of rational roots, thereby expanding its utility beyond merely a method of factorization. By considering the possible combinations of factors of the lead and constant coefficients, one can reverse-engineer polynomials that satisfy certain conditions or fit specific problems.

In summary, the Rational Root Theorem serves as a bridge between the coefficients of a polynomial and its roots, thereby imprinting a deeper understanding of polynomial equations. Its systematic approach to listing potential rational roots enormously simplifies the process of solving polynomials, making it an invaluable tool in algebra.

7.11 Using Synthetic Division to Factor Polynomials

Synthetic division, a streamlined form of polynomial long division, offers a powerful method for dividing polynomials by binomials of the form $x - c$. This technique is particularly useful when factoring polynomials or determining their roots. It simplifies calculations, reduces the possibility of arithmetic errors, and enhances the understanding of polynomial behavior.

The process of synthetic division involves several key steps. To illustrate these steps concisely, let us consider dividing a polynomial $P(x)$ by a binomial $x - c$.

- Identify the coefficient of each term in the polynomial $P(x)$ and write them in descending order of their powers. If any power of x is missing in $P(x)$, represent its coefficient by 0.

- Place the value of c to the left of a vertical bar. The coefficients identified in the previous step are then listed in a row to the right of this bar.

- Draw a horizontal line underneath the row of coefficients. The division process will take place below this line.

- Bring down the leading coefficient to the row below the horizontal line.

- Multiply c by the value just written below the line, and place the result under the next coefficient. Add this result to the coefficient above it, writing the sum below the line.

- Repeat the multiplication and addition steps across the row of coefficients.

Upon completion, the numbers below the line represent the coefficients of the quotient polynomial, except for the last number, which is the remainder. If the remainder is 0, $x - c$ is a factor of the polynomial $P(x)$.

Example:

To further clarify synthetic division, consider the polynomial $P(x) = 2x^3 - 5x^2 + 4x - 1$ and the binomial $x - 2$. We aim to divide $P(x)$ by $x - 2$.

```
Step 1: Coefficients of P(x): 2, -5, 4, -1
Step 2: c = 2
Division setup:
    2 |  2  -5   4  -1
      |_____
      |
```

Executing synthetic division, we proceed as follows:

```
Bring down the 2:
    2 |  2  -5   4  -1
      |_____
      |  2
Multiply and add across:
    2 |  2  -5   4  -1
  x 2 |       4  -2   4
      |_____
      |  2  -1   2   3
```

The resulting row beneath the line reads $2, -1, 2, 3$, indicating the quotient polynomial $2x^2 - x + 2$ and a remainder of 3. Since the remainder is not 0, $x - 2$ is not a factor of $P(x)$.

Applications of Synthetic Division:

Synthetic division is not only used for dividing polynomials but also for finding roots, factoring polynomials, and simplifying the process of polynomial long division. When the remainder is 0, it confirms that $x - c$ is a root of the polynomial, aligning with the Factor Theorem. This capability is extremely beneficial in solving polynomial equations and in the analysis of polynomial functions.

Additionally, synthetic division plays a critical role in the Rational Root Theorem, which states that if a polynomial has rational roots, they can be expressed as fractions $\frac{p}{q}$, where p divides the constant term and q divides the leading coefficient of the polynomial. By testing possible rational roots using synthetic division, one can efficiently determine the actual roots of the polynomial.

To summarize, synthetic division is an indispensable tool in the algebraic toolkit. With practice, it streamlines the division of polynomials, facilitates the discovery of roots, assists in the factoring process, and enhances the overall comprehension of polynomial functions. Mastery of synthetic division thus greatly empowers students to tackle a wide array of problems involving polynomials.

7.12 Factoring Completely

When factoring polynomials, the ultimate goal is to decompose the expression into a product of its irreducible factors over the integers. This process, known as factoring completely, requires the application of several factoring techniques in a strategic sequence. It is essential to employ every tool at our disposal, from identifying the greatest common factor (GCF) to applying special product formulas. This section aims to provide a comprehensive guide on how to factor polynomials completely, ensuring that no further factoring is possible.

Step-by-Step Approach

Factoring completely involves a multi-step approach where each step uses a different factoring technique. The sequence begins with the simplest and most universal methods, gradually moving to more specific cases as the polynomial is simplified further.

1. **Factoring out the GCF:** Start by identifying and factoring out the greatest common factor from all terms of the polynomial. This step simplifies the polynomial and is a crucial foundational move.

2. **Identifying Special Products:** After factoring out the GCF, look for patterns that match special product formulas, such as the difference of squares, the sum or difference of cubes, or perfect square trinomials.

3. **Factoring by Grouping:** For polynomials with four or more terms, factoring by grouping might be effective. This involves grouping terms in a way that each group can be factored separately.

4. **Factoring Trinomials:** At this stage, attempt to factor any trinomials into the product of two binomials. This might require the use of the quadratic formula in cases where the trinomial cannot be easily factored.

5. **Reevaluating Each Factor:** Finally, examine each factor obtained from the previous steps to ensure that it cannot be factored further. Repeat the process for each factor if necessary.

7.12. FACTORING COMPLETELY

Comprehensive Examples

Let us walk through a few examples to demonstrate the process of factoring completely.

Example 1: Factor $12x^4 - 18x^3 - 24x^2$ completely.

1. **Factoring out the GCF:** The GCF of the given polynomial is $6x^2$, so we factor it out.
$$12x^4 - 18x^3 - 24x^2 = 6x^2(2x^2 - 3x - 4)$$

2. **Factoring the Quadratic Expression:** The quadratic expression $2x^2 - 3x - 4$ is then factored into:
$$2x^2 - 3x - 4 = (2x + 1)(x - 4)$$

3. **Final Factored Form:** Combining the factors, we obtain the fully factored expression.
$$12x^4 - 18x^3 - 24x^2 = 6x^2(2x + 1)(x - 4)$$

Example 2: Factor completely $x^4 - 16$.

1. **Identifying Special Products:** The given polynomial is a difference of squares.
$$x^4 - 16 = (x^2)^2 - (4)^2$$

2. **Factoring the Difference of Squares:** Applying the difference of squares formula yields:
$$x^4 - 16 = (x^2 + 4)(x^2 - 4)$$

3. **Continuing to Factor:** Notice that $x^2 - 4$ is also a difference of squares.
$$x^2 - 4 = (x + 2)(x - 2)$$

4. **Final Factored Form:** Combining all factors, we achieve the completely factored form.

$$x^4 - 16 = (x^2 + 4)(x + 2)(x - 2)$$

Practice Problem: To solidify your understanding, try to factor the following polynomial completely: $8x^3 + 27$.

Factoring completely is a vital skill in algebra that requires a deep understanding of various factoring techniques and the ability to apply them in sequence. Mastering this skill enhances problem-solving capabilities and aids in the simplification of complex algebraic expressions. Regular practice with various polynomials ensures proficiency and builds a strong foundation for further studies in algebra and beyond.

Chapter 8

Quadratic Equations and Functions

This chapter delves into quadratic equations and functions, key components of algebra that involve second-degree polynomials. It systematically introduces methods for solving quadratic equations, including factoring, completing the square, and applying the quadratic formula. The chapter further explores the discriminant to determine the nature of the roots of a quadratic equation and investigates the graphing of quadratic functions, highlighting properties such as vertex, axis of symmetry, and parabolas. Through detailed instruction and practice, students will master the techniques for analyzing and solving quadratic equations and functions, enabling them to tackle more complex mathematical problems with confidence.

CHAPTER 8. QUADRATIC EQUATIONS AND FUNCTIONS

8.1 Introduction to Quadratic Equations

Quadratic equations are fundamental elements of algebra, embodying equations of the second degree, which means that the highest power of the unknown variable is two. The general form of a quadratic equation is given by:

$$ax^2 + bx + c = 0$$

where a, b, and c are constants, with $a \neq 0$. The reason a cannot be zero is to ensure the equation remains quadratic, as removing a would reduce the equation to a linear form.

Quadratic equations have been studied over centuries and play a crucial role in various areas of mathematics and science. Solving quadratic equations, therefore, is a key skill in high school algebra that leads to the understanding of more complex concepts in calculus and other advanced areas of mathematics.

The roots or solutions of a quadratic equation can be real or complex and depend entirely on the values of a, b, and c. Finding these roots involves several techniques, ranging from factorization to more advanced algebraic methods. The roots are given by the formula:

$$x = \frac{-b \pm \sqrt{b^2 - 4ac}}{2a}$$

This formula, known as the quadratic formula, provides a straightforward method for solving any quadratic equation, as it directly gives the roots based on the coefficients

8.1. INTRODUCTION TO QUADRATIC EQUATIONS

of the equation.

Another critical aspect of quadratic equations is their graphical representation. Graphically, a quadratic equation represents a parabola, which can either open upwards or downwards, depending on the sign of a. This parabola crosses the y-axis at c, the constant term of the equation. The vertex of the parabola, the point where it turns, offers significant insights into the behavior of the quadratic function.

The importance of understanding the graphical representation of quadratic equations cannot be overstated. It provides a visual understanding of concepts such as the maximum or minimum values a quadratic function can attain, the axis of symmetry which divides the parabola into two mirror images, and the roots of the equation represented by the points where the parabola intersects the x-axis.

In this chapter, we will explore various methods for solving quadratic equations, including factoring, completing the square, and using the quadratic formula. We will also delve into the discriminant, a crucial component of the quadratic formula that helps determine the nature of the roots of the equation. Furthermore, we will investigate the graphing of quadratic functions, understanding the parabola's properties such as vertex and axis of symmetry.

By mastering the techniques for analyzing and solving quadratic equations, students will not only gain confidence in tackling algebraic problems but will also lay the groundwork for exploring more complex mathematical challenges.

```
Example: Solve the quadratic equation 2x^2 - 4x - 6 = 0
```

$$a = 2, \ b = -4, \ c = -6$$

$$x = \frac{-(-4) \pm \sqrt{(-4)^2 - 4(2)(-6)}}{2(2)}$$

$$x = \frac{4 \pm \sqrt{16 + 48}}{4}$$

$$x = \frac{4 \pm \sqrt{64}}{4}$$

$$x = \frac{4 \pm 8}{4}$$

$$x_1 = 3, \ x_2 = -\frac{1}{2}$$

Solution: x = 3, x = -1/2

Through this example, it is evident how the quadratic formula can be used to quickly find the roots of a quadratic equation, underscoring its utility in solving quadratic equations. The subsequent sections will provide further details on each method of solving quadratic equations, followed by their applications in various contexts.

8.2 Solving Quadratic Equations by Factoring

Solving quadratic equations by factoring is among the foundational techniques in algebra. This method is applicable for quadratic equations that can be expressed in the standard form $ax^2 + bx + c = 0$, where a, b, and c are constants. Factoring involves rewriting the quadratic equation as a product of two or more simpler expressions that, when multiplied, yield the original equation. The fundamental principle behind this method is based on the fact

8.2. SOLVING QUADRATIC EQUATIONS BY FACTORING

that if the product of two expressions is zero, then at least one of the expressions must be zero. This section thoroughly explains the process of factoring quadratic equations and employing the Zero Product Property to find the solutions to these equations.

To begin the process, ensure that the quadratic equation is in standard form. If not, manipulate the equation to move all terms to one side, setting the equation equal to zero. This consolidation is crucial for identifying the coefficients a, b, and c, which play a significant role in the factoring process.

The next step involves identifying factors of ac (the product of the coefficient of x^2 and the constant term) that add up to b, the coefficient of x. This is often done through listing pairs of factors of ac and selecting the pair that sums to b. Once the appropriate pair of factors is identified, the quadratic equation is rewritten, splitting the middle term, bx, into two terms whose coefficients are the factors identified for ac.

$$ax^2 + bx + c = ax^2 + (m+n)x + c$$

where m and n are the identified factors of ac that sum to b. The equation is now set for factoring by grouping, which involves dividing the equation into two groups and factoring out the Greatest Common Factor (GCF) from each.

After grouping and factoring the GCF from each group, the equation will generally appear as follows:

$$(a_1x + m_1)(a_2x + m_2) = 0$$

where a_1, m_1, a_2, and m_2 represent constants. At this stage, the Zero Product Property is applied. According to this property, if the product of two expressions equals zero, then at least one of the expressions must be zero. This leads to setting each factor equal to zero:

$$a_1 x + m_1 = 0$$
$$a_2 x + m_2 = 0$$

Solving each of these linear equations for x yields the roots or solutions of the original quadratic equation.

Example:

Consider the quadratic equation $x^2 - 5x + 6 = 0$. The product ac equals 6, and we seek two numbers that multiply to 6 and add up to -5. The numbers -2 and -3 meet these criteria.

$$x^2 - 5x + 6 = (x - 2)(x - 3) = 0$$

Applying the Zero Product Property gives:

$$x - 2 = 0 \text{ or } x - 3 = 0$$

Thus, solving for x in each equation, we find the solutions:

```
x = 2
x = 3
```

To encapsulate, solving quadratic equations by factoring is a critical skill in algebra that hinges on transforming the equation into a product of simpler expressions. This method is particularly effective for equations where factors can be easily identified. Mastery of this technique enhances the mathematical toolkit, enabling more efficient problem-solving and paving the way for further exploration of algebraic concepts.

8.3 Solving Quadratic Equations by Completing the Square

Completing the square is a method used to solve quadratic equations of the form $ax^2 + bx + c = 0$. The principle of this technique is to transform the equation into a perfect square trinomial, thus simplifying the process of finding the roots of the equation. This section outlines the steps for completing the square and applies these steps to solve example equations.

Step 1: If the coefficient of x^2 is not 1, divide every term in the equation by the coefficient of x^2. For example, in $2x^2 + 4x - 6 = 0$, you would divide the entire equation by 2 to get $x^2 + 2x - 3 = 0$.

Step 2: Rearrange the equation by isolating the constant term on one side. Continuing with the simplified equation $x^2 + 2x - 3 = 0$, add 3 to both sides to obtain $x^2 + 2x = 3$.

Step 3: Find a value to complete the square. To do this, take half of the coefficient of x and square it. In our example, the coefficient of x is 2, so half of 2 is 1, and squaring 1 gives us 1. Add this number to both sides of the equation to form a perfect square trinomial on one side. Thus, $x^2 + 2x + 1 = 4$.

Step 4: Factor the perfect square trinomial on the left side of the equation and simplify the right side if necessary. The factored form of $x^2 + 2x + 1$ is $(x+1)^2$, so the equation now reads $(x + 1)^2 = 4$.

Step 5: Solve for x by taking the square root of both sides, remembering to consider both the positive and negative square roots. Continuing with the example, $\sqrt{(x + 1)^2} = \pm\sqrt{4}$, yielding $x + 1 = \pm 2$.

Finally, solve for x by isolating it on one side of the equation. In the example, subtract 1 from both sides to obtain $x = -1 \pm 2$. Therefore, the solutions are $x = 1$ and $x = -3$.

```
Example Solutions:
x = 1
x = -3
```

Example: Solve $3x^2 + 6x - 9 = 0$ by completing the square.

1. $Divide\ by\ 3:\ x^2 + 2x - 3 = 0$
2. $Rearrange:\ x^2 + 2x = 3$
3. $Complete\ the\ square:\ x^2 + 2x + 1 = 4$
4. $Factor\ and\ simplify:\ (x + 1)^2 = 4$
5. $Solve\ for\ x:\ x + 1 = \pm 2$
   $x = -1 \pm 2$

Thus, the solutions to the equation $3x^2 + 6x - 9 = 0$ are $x = 1$ and $x = -3$.

```
Example Solutions:
x = 1
x = -3
```

Completing the square can also be a useful technique for expressing quadratic equations in vertex form, identifying the vertex of a parabola, and analyzing the properties of quadratic functions. This method is especially effective when the quadratic equation does not factor neatly or when applying the quadratic formula is not straightforward. Mastery of this technique enhances problem-solving flexibility in dealing with quadratic equations and functions.

8.4 Solving Quadratic Equations Using the Quadratic Formula

The quadratic formula is a powerful tool for solving equations of the form $ax^2 + bx + c = 0$, where a, b, and c are constants, and $a \neq 0$. This method is particularly useful when the quadratic equation is not easily factorable, making it an indispensable part of our toolkit for dealing with quadratic equations.

The Quadratic Formula

The roots of a quadratic equation can be found using the formula:
$$x = \frac{-b \pm \sqrt{b^2 - 4ac}}{2a}.$$

This formula gives us the x-values at which $ax^2 + bx + c = 0$. The expression under the square root, $b^2 - 4ac$, is known as the discriminant and plays a crucial role in determining the nature of the roots, as discussed in the previous sections.

Derivation of the Quadratic Formula

To derive the quadratic formula, we start with the general form of a quadratic equation:

$$ax^2 + bx + c = 0.$$

Dividing every term by a gives us:

$$x^2 + \frac{b}{a}x + \frac{c}{a} = 0.$$

Rearranging:

$$x^2 + \frac{b}{a}x = -\frac{c}{a}.$$

We complete the square by adding $(\frac{b}{2a})^2$ to both sides:

$$x^2 + \frac{b}{a}x + (\frac{b}{2a})^2 = -\frac{c}{a} + (\frac{b}{2a})^2.$$

Simplifying, we obtain:

$$(x + \frac{b}{2a})^2 = \frac{b^2}{4a^2} - \frac{c}{a}.$$

Taking the square root of both sides, remembering to include both the positive and negative solutions, leads to:

$$x + \frac{b}{2a} = \pm\sqrt{\frac{b^2}{4a^2} - \frac{c}{a}}.$$

Finally, solving for x, we arrive at the quadratic formula:

$$x = \frac{-b \pm \sqrt{b^2 - 4ac}}{2a}.$$

8.4. SOLVING QUADRATIC EQUATIONS USING THE QUADRATIC FORMULA

Application and Examples

Example 1: Solve the quadratic equation $2x^2 - 4x - 6 = 0$. Plugging into the formula, we have $a = 2$, $b = -4$, and $c = -6$:

$$\begin{aligned} x &= \frac{-(-4) \pm \sqrt{(-4)^2 - 4(2)(-6)}}{2(2)} = \frac{4 \pm \sqrt{16 + 48}}{4} \\ &= \frac{4 \pm \sqrt{64}}{4} \\ &= \frac{4 \pm 8}{4} \\ &= 3, -1. \end{aligned}$$

Hence, the equation has two real and distinct solutions $x = 3$ and $x = -1$.

```
Solutions: x = 3, x = -1
```

Example 2: Solve $x^2 + 6x + 9 = 0$ using the quadratic formula.
With $a = 1$, $b = 6$, and $c = 9$:

$$\begin{aligned} x &= \frac{-(6) \pm \sqrt{(6)^2 - 4(1)(9)}}{2(1)} = \frac{-6 \pm \sqrt{36 - 36}}{2} \\ &= \frac{-6 \pm 0}{2} \\ &= -3. \end{aligned}$$

This equation has one real and repeated root $x = -3$.

```
Solution: x = -3
```

Practice Problems

Given the quadratic equations below, solve for x using the quadratic formula:

- $3x^2 - 12x + 9 = 0$
- $x^2 - 5x + 6 = 0$
- $4x^2 + 4x + 1 = 0$
- $2x^2 - 3x - 5 = 0$

In summary, the quadratic formula offers a direct route to finding the roots of any quadratic equation. By embedding this tool deeply into our algebraic arsenal, we equip ourselves to address a broad spectrum of problems involving quadratic equations.

8.5 The Discriminant and Nature of Roots

The concept of the discriminant offers a powerful tool for understanding the nature of the roots of a quadratic equation without necessarily solving the equation explicitly. Recall that a quadratic equation can be expressed in its general form as $ax^2 + bx + c = 0$, where a, b, and c are constants. The discriminant, denoted as D, is given by the expression $D = b^2 - 4ac$. The value of the discriminant provides critical information about the roots of the quadratic equation, specifically regarding their nature and quantity.

8.5. THE DISCRIMINANT AND NATURE OF ROOTS

- If $D > 0$, the quadratic equation has two distinct real roots.

- If $D = 0$, the quadratic equation has exactly one real root, or, more precisely, two real roots that are identically equal.

- If $D < 0$, the quadratic equation has no real roots; instead, it has two complex roots.

To elucidate the use of the discriminant in determining the nature of the roots of a quadratic equation, consider the quadratic formula used to find the roots of any quadratic equation:

$$x = \frac{-b \pm \sqrt{D}}{2a} = \frac{-b \pm \sqrt{b^2 - 4ac}}{2a}.$$

The term under the square root, $b^2 - 4ac$, is the discriminant D. It is evident from the quadratic formula that the discriminant directly influences the roots of the equation. When $D > 0$, the presence of two values for x (given by the $\pm$ sign) indicates two distinct real roots. For $D = 0$, the square root of the discriminant is zero, leading to a single real root. In the case where $D < 0$, the square root of a negative number signifies the presence of complex roots, specifically, a pair of complex conjugates.

To practically apply this concept, consider the following examples:

Example 1: Determine the nature of the roots of the equation $2x^2 - 4x + 1 = 0$.

Here, a = 2, b = -4, and c = 1. Thus, the discriminant $D = (-4)^2 - 4(2)(1) = 8$. Since D > 0, the equation has two distinct real roots.

Example 2: Determine the nature of the roots of the equation $x^2 - 6x + 9 = 0$.

In this case, a = 1, b = -6, and c = 9. Therefore, $D = (-6)^2 - 4(1)(9) = 0$. Since D = 0, the equation has exactly one real root, or two real roots that are equal.

Example 3: Determine the nature of the roots of the equation $x^2 + x + 1 = 0$.

For this equation, a = 1, b = 1, and c = 1. Thus, $D = (1)^2 - 4(1)(1) = -3$. Since D < 0, the equation has two complex roots.

Understanding the discriminant's role in determining the nature of the roots of a quadratic equation is crucial not only for solving these equations but also for predicting the types of solutions that might arise based on the coefficients of the equation. This understanding can be particularly beneficial when the exact roots are not required, or when assessing the feasibility of real solutions in practical applications.

Lastly, it is essential to recognize the impact of the coefficients a, b, and c on the value of the discriminant and, consequently, on the nature of the equation's roots. This observation emphasizes the interconnectedness of the components of the quadratic equation, reinforcing the holistic view of mathematics where each component plays a significant role in the equation's behavior and solutions.

8.6 Applications of Quadratic Equations

Understanding the applications of quadratic equations is pivotal in appreciating their significance beyond the theoretical realm. In this section, we will delve into various real-world scenarios where quadratic equations are not merely abstract mathematical concepts but tools that facilitate problem-solving in fields such as physics, engineering, finance, and economics.

The fundamental form of a quadratic equation is $ax^2 + bx + c = 0$, where a, b, and c are constants, and $a \neq 0$. The solutions to this equation, known as the roots, can be found using methods previously discussed: factoring, completing the square, and the quadratic formula. These solutions often represent measurable quantities in practical applications, making quadratic equations indispensable in many areas.

Projectile Motion in Physics

One of the most common applications of quadratic equations is in the study of projectile motion. When an object is thrown, kicked, or launched into the air, ignoring air resistance, its trajectory can be modeled by a quadratic equation. The object's height above the ground at any time t can be given by $h(t) = -\frac{1}{2}gt^2 + v_0 t + h_0$, where g is the acceleration due to gravity, v_0 is the initial velocity, and h_0 is the initial height. Solving this quadratic equation for t can yield the time at which the object reaches a certain height or the time of impact when it returns to the ground.

Example: A ball is thrown upwards from a height of 1.5 meters with an initial velocity of 10 m/s. Determine the time it takes for the ball to reach the ground.

Optimization Problems in Economics

Quadratic equations often represent cost, revenue, or profit functions in economics, making them essential in determining maximum profit or minimum cost. Given a quadratic function representing cost or revenue, finding the vertex of the parabola can reveal the quantity of production that maximizes profit or minimizes cost.

Dimensions of Rectangular Areas

In geometric problems, quadratic equations can be used to find the dimensions of rectangular plots given constraints. For instance, if the perimeter and the area of a rectangular plot are known, a quadratic equation can be formed to solve for the lengths of the sides.

Example: A rectangular garden has a perimeter of 100 meters, and its area is 600 square meters. Find the dimensions of the garden.

Population Growth Models

Some models of population growth, particularly those that account for a carrying capacity – the maximum sustainable population – lead to quadratic equations. These models are crucial in ecology and environmental science, helping predict population sizes over time and assess the impact on ecosystems.

Financial Mathematics

In finance, quadratic equations can be used to model various scenarios, including calculating profit for businesses given certain costs and revenue structures or determining the break-even points for investments. Additionally, in the context of interest rates and loan repayments, quadratic equations play a crucial role.

To capitalize on the wide range of applications that quadratic equations offer, it is paramount to not only grasp their theoretical aspects but also develop the ability to apply them to solve real-world problems. This application-centric approach not only deepens mathematical understanding but also equips learners with practical skills applicable in numerous fields.

As we explore quadratic equations' diverse applications, it is evident that their practical utility spans across various disciplines. The ability to formulate, solve, and interpret the solutions of quadratic equations in real-world contexts underscores the importance of mastering these algebraic tools. Through these applications, learners can bridge the gap between theoretical mathematics and practical problem-solving, fostering a deeper appreciation for the role of mathematics in understanding and navigating the world.

8.7 Graphing Quadratic Functions

Quadratic functions, defined as functions of the form $f(x) = ax^2 + bx + c$, where a, b, and c are constants and $a \neq 0$, are pivotal in understanding a wide range of algebraic and geometric concepts. This section aims to elu-

CHAPTER 8. QUADRATIC EQUATIONS AND FUNCTIONS

cidate the process and techniques involved in graphing these functions, emphasizing their graphical representation as parabolas.

The graph of a quadratic function is known as a parabola. This U-shaped curve has several important features, including the vertex, the axis of symmetry, and the direction (opening upwards or downwards) that we will explore.

To start with, the vertex of the parabola is a significant point where the graph changes direction. It represents the maximum or minimum point of the function, depending on the direction in which the parabola opens. The coordinates of the vertex can be found using the formula (h, k), where $h = -\frac{b}{2a}$ and $k = f(h)$. It is important to note that when $a > 0$, the parabola opens upwards, and the vertex represents the minimum point. Conversely, if $a < 0$, it opens downwards, and the vertex signifies the maximum point.

The axis of symmetry is a vertical line that passes through the vertex, dividing the parabola into two symmetrical halves. The equation of the axis of symmetry can be given by $x = -\frac{b}{2a}$. This line is instrumental in graphing the parabola as it provides a reference point around which the graph is symmetric.

To graph a quadratic function, one typically begins by identifying the vertex and the axis of symmetry. After plotting these on the coordinate plane, additional points on either side of the axis of symmetry are determined to give the parabola its shape. It is helpful to choose simple values for x to calculate corresponding y values, ensuring that both sides of the parabola are evenly represented.

Let's illustrate this process with an example. Consider

8.7. GRAPHING QUADRATIC FUNCTIONS

the quadratic function $f(x) = x^2 - 4x + 3$.

First, we find the vertex.
$$h = -\frac{-4}{2(1)} = 2, \quad k = (2)^2 - 4(2) + 3 = -1$$
Therefore, the vertex is $(2, -1)$.
The axis of symmetry is $x = 2$.

```
Vertex: (2, -1)
Axis of Symmetry: x = 2
```

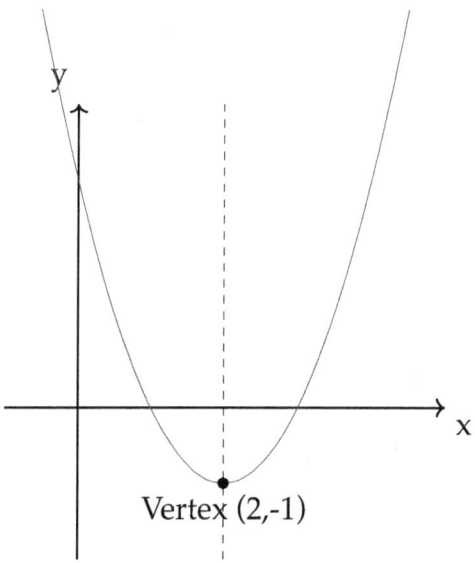

Vertex (2,-1)

The axis of symmetry and several points on either side are plotted to clearly define the shape of the parabola. As can be seen, it opens upwards, signifying that the vertex at $(2, -1)$ represents the minimum point of the function.

Finally, the process of graphing quadratic functions can also benefit from understanding the intercepts. The y-intercept occurs where $x = 0$, and for our example, it is easily found to be $f(0) = 3$. Finding the x-intercepts (or roots) involves solving the equation $x^2 - 4x + 3 = 0$ for x which, depending on the discriminant, may yield real or complex solutions.

Through the use of strategic points like the vertex, axis of symmetry, and intercepts, and the drawing of a smooth curve that captures the essence of the quadratic function, students can efficiently graph quadratic functions. This not only aids in visualizing mathematical relationships but also lays the groundwork for deeper exploration of properties related to parabolas and their applications in various fields.

In summary, the graph of a quadratic function encapsulates critical elements that define its nature and behavior. Understanding these components and the process to graph them allows for a comprehensive grasp of quadratic functions, their properties, and their significance in algebra.

8.8 Properties of Parabolas

A parabola is a U-shaped graph that represents a quadratic function of the form $y = ax^2 + bx + c$, where a, b, and c are constants and $a \neq 0$. The shape and direction of a parabola depend primarily on the value of a. If $a > 0$, the parabola opens upwards, and if $a < 0$, it opens downwards. This section explores the fundamental properties of parabolas that are essential for understanding and graphing quadratic functions.

Vertex

The vertex of a parabola is the highest or lowest point on the graph, depending on the direction the parabola opens. It is the point where the parabola changes direction. The coordinates of the vertex (h, k) can be found using the formula:

$$h = -\frac{b}{2a}, \quad k = c - \frac{b^2}{4a}.$$

Axis of Symmetry

The axis of symmetry is a vertical line that passes through the vertex of the parabola and divides it into two symmetric halves. The equation of the axis of symmetry can be derived from the vertex formula and is given by:

$$x = -\frac{b}{2a}.$$

This property ensures that for any point on the parabola, there exists another point equidistant from the axis of symmetry, having the same y-value but the opposite x-value relative to the axis.

Focus and Directrix

Another significant property of parabolas involves the focus and directrix. The focus is a point F located inside the parabola, and the directrix is a line that lies outside the parabola. Every point P on the parabola is equidistant from the focus and the directrix. If the vertex of the parabola is at (h, k) and the distance between the vertex and the focus (or the vertex and the directrix) is p, then the focus F has coordinates $(h, k+p)$, and the equation of the directrix is $y = k - p$. The value of p is related to the coefficient a in the quadratic equation such that $p = \frac{1}{4a}$ for a parabola opening up or down, and $p = -\frac{1}{4a}$ for a parabola opening left or right.

Width of the Parabola

The width of a parabola refers to the spread of its "arms" as it extends away from the vertex. It is directly proportional to the distance between any two points on the parabola that have the same y-value. The width increases as the absolute value of the coefficient a in the quadratic function decreases. Conversely, the parabola becomes narrower as the absolute value of a increases.

Intercepts

A parabola can intersect the x-axis at zero, one, or two points, known as the x-intercepts or roots of the quadratic equation. These points can be found by setting $y = 0$ in the quadratic equation and solving for x. Similarly, the y-intercept of a parabola is the point at which the graph crosses the y-axis and can be found by setting $x = 0$ in the equation, yielding the point $(0, c)$.

Examples

Consider the parabola described by the quadratic function $y = x^2 - 4x + 3$.

- The vertex can be calculated using $h = -\frac{b}{2a} = 2$ and

$k = c - \frac{b^2}{4a} = -1$. Therefore, the vertex is at $(2, -1)$.

- The axis of symmetry is $x = -\frac{b}{2a} = 2$.

- To find the x-intercepts, we set $y = 0$ and solve the quadratic equation $x^2 - 4x + 3 = 0$ to get $x = 1$ and $x = 3$, hence the x-intercepts are $(1, 0)$ and $(3, 0)$.

- The y-intercept is found by setting $x = 0$, yielding $y = 3$, which gives the point $(0, 3)$.

Understanding these properties allows for a complete analysis of parabolas and provides the tools necessary for graphing quadratic functions accurately. By leveraging these characteristics, one can quickly sketch the graph of a quadratic function, identify its key features, and apply this knowledge to solve real-world problems involving quadratic equations.

8.9 Vertex Form of a Quadratic Function

A quadratic function is typically presented in the standard form $y = ax^2 + bx + c$. However, for various analytical purposes such as identifying the vertex or transforming the graph, the vertex form ($y = a(x - h)^2 + k$) offers a more direct approach. In this section, we delineate the process of converting a quadratic function from its standard form to its vertex form, elaborating on the implications and utility of this transformation.

Firstly, it is paramount to understand the components of the vertex form. In the equation $y = a(x - h)^2 + k$, the point (h, k) represents the vertex of the parabola, and the value of a determines the direction and "width" of the parabola. If $a > 0$, the parabola opens upwards; if $a < 0$, it opens downwards. The "width" of the parabola is inversely related to the absolute value of a; larger values of $|a|$ result in a "narrower" parabola, while smaller values of $|a|$ yield a "wider" parabola.

Converting from Standard to Vertex Form.

The process of rewriting a quadratic function in the vertex form involves completing the square. This technique modifies the quadratic portion of the standard form equation into a perfect square trinomial, making the function easier to analyze and graph.

Consider the quadratic equation in standard form:

$$y = ax^2 + bx + c$$

To convert this equation into its vertex form, follow these steps:

1. Isolate the quadratic and linear terms on one side by subtracting c from both sides, resulting in $y - c = ax^2 + bx$.

2. Factor out the coefficient a from the right-hand side, giving $y - c = a(x^2 + \frac{b}{a}x)$.

3. Complete the square for the expression within the parentheses. This involves adding and subtracting $\left(\frac{b}{2a}\right)^2$, the square of half the coefficient of x, to the right-hand side.

4. This step yields $y - c + a\left(\frac{b}{2a}\right)^2 = a\left(x^2 + \frac{b}{a}x + \left(\frac{b}{2a}\right)^2\right) - a\left(\frac{b}{2a}\right)^2$.

8.9. VERTEX FORM OF A QUADRATIC FUNCTION

5. Simplify the expression by combining like terms and repositioning to achieve the vertex form, resulting in the final expression:

$$y = a(x - \frac{b}{2a})^2 + (c - a\left(\frac{b}{2a}\right)^2)$$

Here, the vertex (h, k) of the parabola can be directly identified as $\left(\frac{b}{2a}, c - a\left(\frac{b}{2a}\right)^2\right)$.

Example: Converting from Standard to Vertex Form

Given the quadratic function $y = 2x^2 + 8x + 3$, convert it into its vertex form.

1. Start with $y = 2x^2 + 8x + 3$.

2. Following the steps outlined, we subtract 3 from both sides and factor out the leading coefficient from the right-hand side, yielding $y - 3 = 2(x^2 + 4x)$.

3. Next, complete the square by adding and subtracting 2^2 (inside the parenthesis, after factoring out 2), leading to $y - 3 = 2(x^2 + 4x + 4 - 4)$.

4. The equation simplifies to $y - 3 + 8 = 2(x + 2)^2 - 8$ and finally to $y = 2(x + 2)^2 + 5$.

The transformed equation in vertex form is $y = 2(x+2)^2 + 5$, with the vertex of the parabola being $(-2, 5)$.

```
Vertex: (-2, 5)
```

Understanding the conversion to vertex form offers profound implications in graphing quadratic functions, optimizing equations, and solving real-world problems efficiently. Mastery of this technique empowers students to navigate the complexities of quadratic functions with improved intuition and analytical skills.

8.10 Transformations of Quadratic Functions

Quadratic functions, characterized by the general form $y = ax^2 + bx + c$, where a, b, and c are constants, are fundamental to understanding a wide range of algebraic and geometric concepts. A basic understanding of quadratic functions lays the groundwork for exploring their transformations, which include translations, reflections, stretches, and compressions. These transformations alter the appearance of the graph of the quadratic function but do so in predictable and understandable ways.

To begin, consider the vertex form of a quadratic function: $y = a(x - h)^2 + k$. This form explicitly reveals the vertex of the parabola as (h, k). The parameters h and k allow for horizontal and vertical translations of the parabola, respectively.

Horizontal Translation: A change in the value of h results in a horizontal translation. If h is positive, the parabola moves h units to the right; if h is negative, it moves $|h|$ units to the left. This can be conceptually understood as moving the entire graph along the x-axis without altering its shape.

Vertical Translation: Similarly, altering the value of k translates the parabola vertically. A positive k lifts the parabola k units upwards, while a negative k lowers it $|k|$ units. This translation is equivalent to moving the graph

8.10. TRANSFORMATIONS OF QUADRATIC FUNCTIONS

up or down the y-axis.

Reflection: A reflection across the x-axis is achieved by negating the coefficient a in the vertex form. If a is positive, the parabola opens upwards; if a is negative, the parabola reflects across the x-axis to open downwards. This modification does not change the parabola's vertex but alters its direction.

Vertical Stretch and Compression: The absolute value of a impacts the vertical stretch or compression of the parabola. If $|a| > 1$, the parabola becomes narrower, indicating a vertical stretch. Conversely, if $0 < |a| < 1$, the graph widens, suggesting a vertical compression. This effect is due to the increase or decrease in the rate at which the y-value changes relative to the x-value.

Practical Example:

Consider the transformation of the basic quadratic function $y = x^2$ to $y = 2(x - 3)^2 + 4$. Here, $a = 2$ indicates a vertical stretch since $|2| > 1$. The function is also translated 3 units to the right and 4 units up due to the values of $h = 3$ and $k = 4$.

```
Original Function: y = x^2
Transformed Function: y = 2(x - 3)^2 + 4
```

To visually comprehend these transformations, plotting both the original and transformed functions is insightful.

Transformation of Quadratic Functions

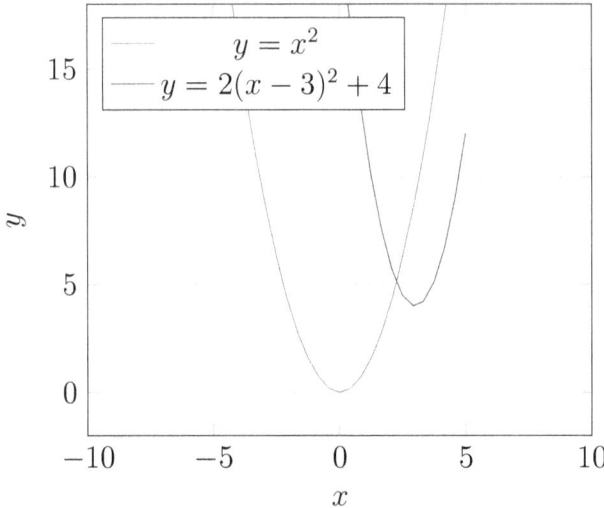

In summary, transformations of quadratic functions enable the manipulation of the basic parabolic shape in multiple ways: by shifting its position through translations, flipping it through reflections, and altering its width through stretches and compressions. Each of these transformations applies a precise mathematical change to the function's equation, resulting in a corresponding visual change to its graph. Understanding these transformations allows for the construction and analysis of quadratic functions in a variety of contexts, from solving real-world problems to exploring advanced mathematical theories.

- Horizontal translations are controlled by the h value in the vertex form, moving the graph left or right.

- Vertical translations depend on the k value, shifting the graph up or down.

- Reflections are achieved by negating the a value, flipping the graph across the x-axis.

- Vertical stretches and compressions are dictated by the magnitude of a, affecting the graph's width and steepness.

Through diligent practice and analysis, one can master the intricacies of these transformations, greatly enhancing their understanding and application of quadratic functions.

8.11 Quadratic Inequalities

In this section, we shift our focus towards understanding and solving quadratic inequalities. A quadratic inequality is similar to a quadratic equation; however, instead of an equality, it involves an inequality symbol ($<, \leq, >, \geq$). The standard form of a quadratic inequality is expressed as $ax^2 + bx + c < 0$, $ax^2 + bx + c \leq 0$, $ax^2 + bx + c > 0$, or $ax^2 + bx + c \geq 0$, where $a \neq 0$. Solving quadratic inequalities is a pivotal skill in algebra that aids in understanding a wide range of mathematical problems, from simple algebraic questions to complex calculus applications.

The first step in solving a quadratic inequality is to find the roots of the corresponding quadratic equation $ax^2 + bx + c = 0$. These roots can be found using methods such as factoring, completing the square, or applying the quadratic formula. Once the roots are identified, they are used to divide the number line into intervals. The next step involves selecting test points from each interval to substitute into the original inequality to determine which intervals satisfy the inequality.

Let us begin with an example to illustrate this process in detail.

Example: Solve the quadratic inequality $x^2 - 3x - 4 < 0$.

Step 1: Identify the corresponding quadratic equation and find its roots.

$$x^2 - 3x - 4 = 0$$
$$(x-4)(x+1) = 0$$

Hence, the roots of the equation are $x = 4$ and $x = -1$. These roots divide the number line into three intervals: $(-\infty, -1)$, $(-1, 4)$, and $(4, \infty)$.

Step 2: Choose test points from each interval and substitute them into the original inequality. Suitable test points would be $x = -2$, $x = 0$, and $x = 5$.

- Testing $x = -2$ in $x^2 - 3x - 4 < 0$ gives $4 - 6 - 4 < 0$, which simplifies to $-6 < 0$. This means the interval $(-\infty, -1)$ satisfies the inequality.

- Testing $x = 0$ yields $0 - 0 - 4 < 0$, which also satisfies the inequality. Thus, the interval $(-1, 4)$ satisfies the inequality.

- Testing $x = 5$ gives $25 - 15 - 4 < 0$, which is not true. Therefore, the interval $(4, \infty)$ does not satisfy the inequality.

From the test points and intervals assessed, the solution to the inequality $x^2 - 3x - 4 < 0$ is $x \in (-\infty, -1) \cup (-1, 4)$.

The graphical approach to solving quadratic inequalities provides a visual understanding of the solution. By

8.11. QUADRATIC INEQUALITIES

graphing the corresponding quadratic equation $y = ax^2 + bx + c$, the solution to the inequality can be visualized as regions where the graph lies above or below the x-axis, depending on the inequality's direction.

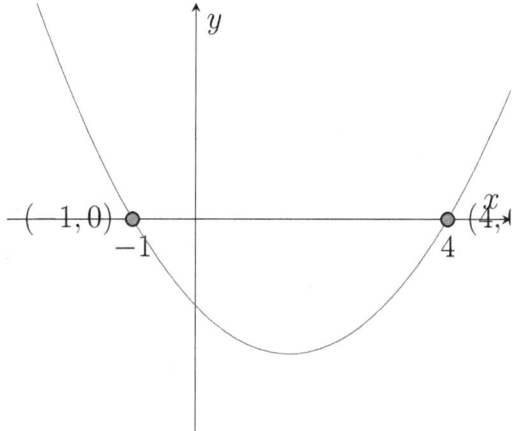

In the graph, it is evident that the curve intersects the x-axis at $x = -1$ and $x = 4$, dividing the x-axis into three segments. The segments $(-\infty, -1)$ and $(-1, 4)$ correspond to the regions where the graph lies below the x-axis, representing the solution to the inequality $x^2 - 3x - 4 < 0$.

To solve quadratic inequalities is to understand the role of the roots of the quadratic equation and to effectively analyze the intervals they create on the number line. Through the utility of test points and graphical representation, one can visually and algebraically determine the set of values that satisfy the inequality.

As one moves forward with solving more complex inequalities, it becomes essential to grasp the underlying principles discussed in this section. Mastery of these concepts facilitates the exploration of further mathematical theories and applications, demonstrating the beauty and interconnectedness of mathematical ideas.

8.12 The Quadratic Formula and the Discriminant

In this section, we shall delve into the quintessence of solving quadratic equations: the Quadratic Formula. This formidable tool offers a direct method for finding the roots of any quadratic equation, with $ax^2 + bx + c = 0$, where a, b, and c are constants, and $a \neq 0$. Let us then proceed to explore the concept of discriminant, which reveals the nature of the roots that a quadratic equation possesses.

The Quadratic Formula

To derive the quadratic formula, we start with the general form of a quadratic equation:

$$ax^2 + bx + c = 0.$$

Through completing the square, this equation can be rewritten in the form of:

$$x = \frac{-b \pm \sqrt{b^2 - 4ac}}{2a}.$$

This expression is known as the Quadratic Formula. The symbol $\pm$ signifies that there are typically two solutions to the equation: one with a positive root and one with a negative root of the same magnitude.

8.12. THE QUADRATIC FORMULA AND THE DISCRIMINANT

Examples of Applying the Quadratic Formula

Let us consider the quadratic equation $2x^2 - 4x - 6 = 0$. Applying the quadratic formula yields:

```
x = 2, x = -1.5
```

Now, let us evaluate a scenario where the quadratic equation does not have real solutions. For instance, $x^2 + 4x + 5 = 0$ results in:

```
x = -2 + i, x = -2 - i
```

This indicates that the solutions are complex numbers.

The Discriminant

The term under the square root in the quadratic formula, $b^2 - 4ac$, is called the discriminant and is denoted by D. The value of the discriminant tells us the nature of the roots of the quadratic equation. Specifically:

- If $D > 0$, the quadratic equation has two distinct real roots.

- If $D = 0$, the quadratic equation has exactly one real root (also known as a repeated root).

- If $D < 0$, the quadratic equation has two complex roots.

The discriminant offers a quick way to determine the type of solutions without explicitly solving the quadratic equation.

Example Utilizing the Discriminant

Consider the quadratic equation $3x^2 - 6x + 2 = 0$. To determine the nature of its roots, calculate the discriminant:

```
D = (-6)^2 - 4(3)(2) = 12,
```

Since $D > 0$, the equation has two distinct real roots. Thus, before even applying the quadratic formula, we understand the qualitative nature of its solutions.

Graphical Interpretation of the Discriminant

The value of the discriminant also has a graphical interpretation in terms of the quadratic function's graph. This interpretation is based on the interaction between the parabola and the x-axis:

- If $D > 0$, the graph of the quadratic function intersects the x-axis at two points, corresponding to the two distinct real roots.

- If $D = 0$, the graph of the quadratic function touches the x-axis at exactly one point, indicating a repeated root.

- If $D < 0$, the graph of the quadratic function does not intersect the x-axis, indicating complex roots.

By integrating the quadratic formula and the concept of discriminant, students gain a powerful toolkit for solving quadratic equations and understanding the nature of their solutions. These concepts not only underscore the utility of quadratic equations in mathematics but also exemplify the elegance and coherence of algebraic theory.

Chapter 9

Radical Expressions and Equations

This chapter introduces radical expressions and equations, focusing on concepts involving roots and powers. It covers the simplification of radical expressions, operations with radicals including addition, subtraction, multiplication, and division, and the process of rationalizing the denominator. The discussion extends to solving equations that contain radical expressions, utilizing graphical methods for understanding radical functions, and exploring the implications of complex numbers. By equipping students with strategies to manipulate and solve radical expressions and equations, this chapter prepares them for further studies in algebra and related fields, emphasizing the importance of radicals in various mathematical contexts.

CHAPTER 9. RADICAL EXPRESSIONS AND EQUATIONS

9.1 Introduction to Radical Expressions

Radical expressions form a fundamental part of algebra and play a vital role in various mathematical operations and problem-solving. A radical expression is any expression that includes a radical symbol, $\sqrt{}$, with the radical sign denoting the square root of the number or expression underneath it. More generally, radicals can also involve nth roots, represented as $\sqrt[n]{}$, where n indicates the degree of the root.

To fully comprehend radical expressions, it is essential to understand the concept of roots and powers. Powers, also known as exponents, involve raising a base number to a certain power. Conversely, a root operation aims to identify a number, which, when raised to a specific power, yields the original number under the radical sign. For instance, in the expression $\sqrt{9}$, we seek a number that, when squared, equals 9. The answer is 3, as $3^2 = 9$. Therefore, $\sqrt{9} = 3$.

In the context of nth roots, the expression $\sqrt[n]{x}$ asks for a number that, when raised to the nth power, equals x. This is succinctly captured by the equation $y^n = x$, implying $\sqrt[n]{x} = y$. For example, $\sqrt[3]{8} = 2$ because $2^3 = 8$.

Let us delve deeper into the properties and manipulation of radical expressions. One of the first aspects to consider is the simplification of radicals. Simplifying a radical involves finding an equivalent expression that is as expressive and straightforward as possible. This often includes identifying and extracting perfect nth powers from under the radical sign. For example, $\sqrt{50}$ can be simplified to $5\sqrt{2}$ because $50 = 25 \times 2$ and $\sqrt{25} = 5$.

9.1. INTRODUCTION TO RADICAL EXPRESSIONS

Radical expressions can also be subjected to arithmetic operations such as addition, subtraction, multiplication, and division. However, it's important to note that addition and subtraction of radicals can only be performed if the radicals have the same radicand (the number under the radical sign) and index (the degree of the root). For instance, $\sqrt{3} + 2\sqrt{3} = 3\sqrt{3}$, while $\sqrt[3]{2} + \sqrt{2}$ cannot be simplified in a similar manner due to differing radicands and indices.

On the other hand, radical expressions can always be multiplied and divided, regardless of the radicand and index, by employing properties of radicals. A key property is that the product (or quotient) of two radicals with the same index is equivalent to the radical of the product (or quotient) of the radicands, as shown by $\sqrt{a}\sqrt{b} = \sqrt{ab}$ and $\frac{\sqrt{a}}{\sqrt{b}} = \sqrt{\frac{a}{b}}$, provided $b \neq 0$ for division.

Rationalizing the denominator is another crucial technique in dealing with radical expressions, especially in expressions of the form $\frac{a}{\sqrt{b}}$. The goal is to eliminate the radical from the denominator by multiplying both the numerator and denominator by an appropriate expression that clears the radical.

Conclusively, radical expressions embody a rich and intricate area of algebra that bridges the conceptual gap between powers and roots. Mastery of radical expressions and their properties lays the groundwork for tackling more complex algebraic equations and functions, reinforcing the significance of radicals in the broader landscape of mathematical inquiry.

9.2 Simplifying Radical Expressions

Understanding how to simplify radical expressions is a fundamental skill in algebra that allows students to manipulate and evaluate expressions involving roots more effectively. This section will guide you through the essential principles and methods for simplifying radical expressions, ensuring a solid foundation before advancing to more complex operations with radicals.

A radical expression contains a radical sign ($\sqrt{}$) with an expression underneath it, known as the radicand. The radical sign can denote not only a square root but also higher roots such as cube roots ($\sqrt[3]{}$), fourth roots ($\sqrt[4]{}$), and so on. The degree of the root—indicating whether it is a square root, cube root, or another root—is shown by a small number outside and to the left of the radical sign. If no number is displayed, the root is assumed to be 2 (a square root).

Principle of Simplifying Radical Expressions

The primary goal when simplifying radical expressions is to transform the expression into the simplest possible form. This involves:

- Ensuring that the radicand (the number or expression under the radical sign) has no perfect square (or cube, fourth, etc., depending on the root) factors besides 1.

- Ensuring that no fractions are present under the radical sign.

- Ensuring that no radicals appear in the denominator of a fraction.

Step-by-Step Approach

Step 1: Factorize the Radicand

Start by factorizing the radicand into its prime factors. For square roots, identify pairs of identical factors; for cube roots, identify sets of three identical factors, and so on. These pairs or sets can be moved outside the radical as single factors.

$$\begin{aligned}\sqrt{48} &= \sqrt{2^4 \times 3} \\ &= \sqrt{(2^2)^2 \times 3} \\ &= 2^2 \times \sqrt{3} \\ &= 4\sqrt{3}\end{aligned}$$

Step 2: Simplify Fractional Radicands

If the radicand is a fraction, attempt to separately root the numerator and the denominator. Simplify each as much as possible.

$$\begin{aligned}\sqrt{\frac{16}{25}} &= \frac{\sqrt{16}}{\sqrt{25}} \\ &= \frac{4}{5}\end{aligned}$$

Step 3: Rationalize the Denominator

If a radical expression exists in the denominator of a fraction, multiply both the numerator and denominator by a form of 1 that will eliminate the radical in the denominator.

$$\frac{1}{\sqrt{3}} = \frac{1}{\sqrt{3}} \cdot \frac{\sqrt{3}}{\sqrt{3}}$$
$$= \frac{\sqrt{3}}{3}$$

Example Problems

Example 1: Simplify the following radical expression: $\sqrt{32}$.

Solution:

1. Factorize the radicand: $32 = 2^5$

2. Identify and move out pairs of factors:
$\sqrt{(32)} = \sqrt{(2^5)} = \sqrt{((2^2)^2 * 2)} = 2^2 * \sqrt{(2)} = 4\sqrt{(2)}$

Final Answer: $4\sqrt{(2)}$

Example 2: Simplify the following expression: $\sqrt{\frac{81}{49}}$.

Solution:

1. Apply square root separately to numerator and denominator: $\sqrt{(81/49)} = \sqrt{(81)}/\sqrt{(49)}$

2. Simplify: $\sqrt{(81)}/\sqrt{(49)} = 9/7$

Final Answer: 9/7

The capability to simplify radical expressions effectively prepares students for advanced topics in algebra, including solving radical equations and understanding the properties of radical functions. Mastering the foundational steps of factorizing the radicand, simplifying

9.3. OPERATIONS WITH RADICAL EXPRESSIONS

fractional radicands, and rationalizing the denominator is essential. Through diligent practice, the simplification process becomes more intuitive, enabling students to tackle more complex algebraic challenges with confidence.

9.3 Operations with Radical Expressions

Understanding operations with radical expressions is fundamental for mastering the concepts in algebra and preparing for more advanced topics. This section will explore the four main operations: addition, subtraction, multiplication, and division of radical expressions.

Addition and Subtraction of Radical Expressions

When adding or subtracting radical expressions, it is crucial to remember that only like radicals can be combined directly. Like radicals have the same index and the same radicand. The process is similar to combining like terms in algebraic expressions.

Consider two radical expressions $\sqrt{a}$ and $\sqrt{b}$. If $a = b$, these expressions are like radicals and can be added or subtracted. The sum or difference of $n\sqrt{a} + m\sqrt{a}$ is $(n + m)\sqrt{a}$, where n and m are coefficients of the radicals.

Example 1:

$$3\sqrt{5} + 2\sqrt{5} = (3 + 2)\sqrt{5} = 5\sqrt{5}$$

Example 2:

$$7\sqrt{3} - 3\sqrt{3} = (7-3)\sqrt{3} = 4\sqrt{3}$$

For radicals with different radicands or indices, simplification or transformation might be necessary to determine if they are like radicals.

Multiplication of Radical Expressions

The multiplication of radical expressions utilizes the property $\sqrt{a} \cdot \sqrt{b} = \sqrt{ab}$, irrespective of whether a and b are like radicals. This property stems from the definition of radicals and the properties of exponents.

Example 3:
$$\sqrt{2} \cdot \sqrt{3} = \sqrt{2 \cdot 3} = \sqrt{6}$$

Multiplication can also involve distributing a radical across a polynomial inside the radical.

Example 4:
$$2\sqrt{3} \cdot 3\sqrt{5} = 6\sqrt{3 \cdot 5} = 6\sqrt{15}$$

Division of Radical Expressions

Division involving radicals employs the concept of rationalizing the denominator, ensuring the denominator of a fraction is rational. A common approach is to multiply both the numerator and denominator by a suitable radical that will eliminate the radical in the denominator.

Given a fraction $\frac{\sqrt{a}}{\sqrt{b}}$, it can be simplified to $\sqrt{\frac{a}{b}}$, provided $b \neq 0$.

9.3. OPERATIONS WITH RADICAL EXPRESSIONS

Example 5:
$$\frac{\sqrt{8}}{\sqrt{2}} = \sqrt{\frac{8}{2}} = \sqrt{4} = 2$$

In cases where the denominator is a binomial involving radicals, multiply the numerator and denominator by the conjugate of the denominator.

Example 6:

$\dfrac{1}{1+\sqrt{2}}$ is rationalized by multiplying by $\dfrac{1-\sqrt{2}}{1-\sqrt{2}}$

$$= \frac{1-\sqrt{2}}{1-2} = \frac{1-\sqrt{2}}{-1} = \sqrt{2}-1$$

Rationalizing Radical Expressions

Rationalizing the denominator is a key operation in dealing with divisions of radical expressions. It involves removing the radical from the denominator by multiplying both the numerator and the denominator by an appropriate expression.

For a simple radical in the denominator, multiply both the numerator and the denominator by the radical in the denominator.

Example 7:

$\dfrac{5}{\sqrt{3}}$ is rationalized by multiplying by $\dfrac{\sqrt{3}}{\sqrt{3}}$

$$= \frac{5\sqrt{3}}{3}$$

For a binomial denominator involving a radical, use its conjugate to multiply both the numerator and the denominator.

Overall, operations with radical expressions require attention to the properties of radicals, combining like terms, and rationalizing denominators for division. Through practice and application of these principles, students can gain proficiency in handling radical expressions, laying a strong foundation for further studies in algebra.

9.4 Rationalizing the Denominator

Rationalizing the denominator is the process of eliminating the radical expression in the denominator of a fraction. This practice is essential for simplifying radical expressions and is a key skill in algebra. It aids in the comparison, addition, subtraction, and division of radical expressions by converting them into an equivalent form that is often easier to understand and manipulate. We will explore methods for rationalizing the denominator, focusing on both monomial and binomial denominators that contain radical expressions.

For monomial denominators, the process involves multiplying both the numerator and the denominator by a suitable radical expression that will eliminate the radical in the denominator. Consider a fraction $\frac{a}{\sqrt{b}}$, where a and b are integers and b is not a perfect square. To rationalize the denominator, we multiply both the numerator and the denominator by $\sqrt{b}$, leading to $\frac{a\sqrt{b}}{b}$.

$$\frac{a}{\sqrt{b}} \times \frac{\sqrt{b}}{\sqrt{b}} = \frac{a\sqrt{b}}{b}$$

9.4. RATIONALIZING THE DENOMINATOR

This process alters the appearance of the fraction but not its value, as multiplying by $\frac{\sqrt{b}}{\sqrt{b}}$ is equivalent to multiplying by 1.

When the denominator is a binomial containing a radical, we use the conjugate of the denominator to rationalize. The conjugate of a binomial $a + \sqrt{b}$ is $a - \sqrt{b}$, and vice versa. Multiplying the denominator by its conjugate eliminates the radical term, owing to the difference of squares identity, which states that $(a+b)(a-b) = a^2 - b^2$.

Consider a fraction $\frac{c}{a+\sqrt{b}}$, where a, b, and c are integers, and b is not a perfect square. To rationalize the denominator, we multiply the fraction by $\frac{a-\sqrt{b}}{a-\sqrt{b}}$:

$$\frac{c}{a+\sqrt{b}} \times \frac{a-\sqrt{b}}{a-\sqrt{b}} = \frac{c(a-\sqrt{b})}{a^2-b}$$

Below is an example illustrating the rationalization of a denominator involving a binomial with a radical.

Example:

Rationalize the denominator of $\frac{3}{2+\sqrt{5}}$.

Solution:

$$\frac{3}{2+\sqrt{5}} \times \frac{2-\sqrt{5}}{2-\sqrt{5}} = \frac{3(2-\sqrt{5})}{(2+\sqrt{5})(2-\sqrt{5})}$$
$$= \frac{3(2-\sqrt{5})}{4-5}$$
$$= \frac{3(2-\sqrt{5})}{-1}$$
$$= -3(2-\sqrt{5})$$

= -6 + 3sqrt(5)

This example demonstrates the utility of the conjugate in eliminating the radical from the denominator. By applying the difference of squares formula, we not only rationalize the denominator but also simplify the expression into a form that can be easily worked with.

It is important for students to master the process of rationalizing the denominator as it frequently appears in algebra, calculus, and other areas of mathematics. This skill is not only about adhering to mathematical convention but also about simplifying expressions for easier manipulation and understanding. Through practice, students will develop the ability to quickly identify the most efficient method to rationalize denominators, whether they involve monomial or binomial radicals.

In summary, rationalizing the denominator is a critical technique in algebra that aids in the simplification of radical expressions. By multiplying both the numerator and the denominator by an appropriate expression, we can eliminate radicals from the denominator, thereby rendering the expression in a form that is often easier to handle and interpret. Mastery of this technique is essential for success in algebra and beyond.

9.5 Radical Equations

Radical equations are equations in which the variable is under a radical, most commonly a square root. Solving radical equations involves isolating the radical on one side of the equation and then eliminating the radical by raising both sides of the equation to the power that corresponds to the root of the radical. This section will guide students through the process of solving radical equations,

9.5. RADICAL EQUATIONS

emphasizing the importance of checking all solutions due to the potential introduction of extraneous solutions.

To commence, consider a basic form of a radical equation:

$$\sqrt{x} = a,$$

where a is a real number. To solve for x, square both sides of the equation to obtain $x = a^2$. It's crucial to note that when we square both sides of an equation, we may introduce solutions that do not satisfy the original equation. Hence, verifying each solution by substituting it back into the original equation is essential.

Example 1. Solve the radical equation $\sqrt{x+4} = 3$.

To solve:

$$\sqrt{x+4} = 3$$
$$(x+4) = 3^2$$
$$x + 4 = 9$$
$$x = 5$$

Checking:

$\sqrt{(5+4)} = \sqrt{9} = 3$

The solution $x = 5$ satisfies the original equation.

Moving forward, for equations involving radicals other than square roots, the principle remains the same—raise both sides of the equation to a power that matches the index of the radical. For instance, to solve a cube root equation $\sqrt[3]{x} = a$, cube both sides to eliminate the radical.

Radical equations may also have the radical term multiplied by a coefficient. In such cases, isolate the radical before proceeding to eliminate the radical. Above all, exercise caution with each step to ensure the integrity of the equation is maintained.

CHAPTER 9. RADICAL EXPRESSIONS AND EQUATIONS

Example 2. Solve the radical equation $2\sqrt{x} - 5 = 3$.

First, isolate the radical:

$$2\sqrt{x} - 5 + 5 = 3 + 5$$
$$2\sqrt{x} = 8$$
$$\sqrt{x} = 4$$

Next, square both sides to eliminate the square root:

$$(\sqrt{x})^2 = 4^2$$
$$x = 16$$

Checking:

2√(16) - 5 = 2×4 - 5 = 3

The solution $x = 16$ is valid.

One common challenge in solving radical equations is the occurrence of extraneous solutions—solutions that emerge from the algebraic process but do not satisfy the original equation. This phenomenon underscores the imperative of substituting potential solutions back into the initial equation to validate them.

Example 3. Solve the radical equation $\sqrt{x-1} + 2 = x$.

Isolate the radical:

$$\sqrt{x-1} + 2 - 2 = x - 2$$
$$\sqrt{x-1} = x - 2$$

Square both sides:

$$(\sqrt{x-1})^2 = (x-2)^2$$
$$x - 1 = x^2 - 4x + 4$$

Rearrange and simplify:
$$0 = x^2 - 5x + 5$$

Solving this quadratic equation might involve methods such as factoring, completing the square, or applying the quadratic formula, resulting in potential solutions. Each solution must then be checked in the original equation to scrutinize its validity.

To bolster understanding, assignments at this point will span a spectrum from straightforward radical equations to those necessitating multiple steps, involving distribution, and requiring the synthesis of the quadratic formula, ensuring students are equipped to tackle radical equations in diverse contexts.

In real-life applications, radical equations can model scenarios where the relationship between variables involves roots, such as in physics for motion under uniform acceleration or in geometry for the formulas involving the dimensions of shapes. Practicing with these equations not only builds algebraic skill but also enhances problem-solving abilities in scientific and engineering contexts.

9.6 Solving Radical Equations

Radical equations are equations in which the variable is located under a radical, typically a square root. Solving radical equations requires isolating the radical expression and then eliminating the radical by raising both sides of the equation to the power that corresponds to the radical. The process occasionally introduces extraneous solutions, thus requiring verification of potential solutions by substituting them back into the original equation. This section details strategies to address radical equations and provides examples to illustrate the methodology.

First, let us solidify the understanding that a radical equation may involve one or more radical terms. The general form can be represented as $\sqrt[n]{x+a} = b$, where n is the index of the root, and a and b are constants. The goal is to find the value of x that satisfies this equation.

Step 1: Isolate the radical expression. To begin solving, the first step is always to isolate the radical on one side of the equation. If there are multiple radicals, focus on isolating one of them first. Consider the equation $\sqrt{x-1} + 2 = 5$. Subtract 2 from both sides to isolate the radical, obtaining $\sqrt{x-1} = 3$.

$$\sqrt{x-1} = 3$$

Step 2: Eliminate the radical. Once the radical is isolated, eliminate it by raising both sides of the equation to the power of the index of the radical. In most high school algebra problems, this will be a square root, so you'll square both sides. From the equation above, squaring both sides gives us $x - 1 = 9$.

$$x - 1 = 9$$

Step 3: Solve for the variable. With the radical eliminated, solve the remaining equation for x. Following our example, add 1 to both sides to find $x = 10$.

x = 10

Step 4: Check for extraneous solutions. It's crucial to substitute the solution back into the original equation to

9.6. SOLVING RADICAL EQUATIONS

ensure it's not extraneous. An extraneous solution arises when the process of squaring both sides introduces valid algebraic solutions that don't satisfy the original equation. For our example, substituting $x = 10$ into $\sqrt{x-1} + 2 = 5$ confirms that 10 is a valid solution.

$\sqrt{10-1} + 2 = 5$ is true.

Multiple Radicals: When an equation contains more than one radical, isolate one radical first, remove it as described, then isolate the other radical and repeat the process. Consider $\sqrt{x} + \sqrt{x-4} = 4$. Isolate one radical, square both sides, simplify, and then isolate the second radical. This process may need to be repeated several times, and always conclude with a check for extraneous solutions.

Example Problem: Solve $\sqrt{2x+3} - \sqrt{x-1} = 1$.

First, isolate one of the radicals:

$$\sqrt{2x+3} = 1 + \sqrt{x-1}$$

Next, square both sides to eliminate the radical on the left:

$$2x + 3 = 1 + 2\sqrt{x-1} + x - 1$$

Then, isolate the remaining radical and square both sides again:

$$2x + 2 = 2\sqrt{x-1}$$
$$x + 1 = \sqrt{x-1}$$

Squaring a second time yields:

$$x^2 + 2x + 1 = x - 1$$
$$x^2 + x + 2 = 0$$

Solving this quadratic equation gives us the potential solutions, which must be checked to ensure they do not introduce any contradictions in the original equation.

This methodical approach to solving radical equations—isolating the radical, eliminating the radical, solving the resulting equation, and checking for extraneous solutions—ensures accurate resolution of these problems. Practice with varied problems will cement these principles and improve problem-solving efficiency.

9.7 Complex Numbers

In this section, we explore the domain of complex numbers, expanding our mathematical landscape beyond the realm of real numbers. Complex numbers are essential for solving equations that have no real solution, particularly those involving square roots of negative numbers. A complex number is expressed in the form $a + bi$, where a and b are real numbers, and i is the imaginary unit with the property that $i^2 = -1$.

9.7. COMPLEX NUMBERS

Definition and Representation

The foundation of complex numbers lies in the imaginary unit i. The importance of i stems from its property, $i^2 = -1$, which cannot be satisfied by any real number. A complex number combines this imaginary unit with a real part to form $a + bi$, where a represents the real part and bi represents the imaginary part.

For instance, in the complex number $3 + 4i$, 3 is the real part and $4i$ is the imaginary part. It is essential to note that while i is defined as $\sqrt{-1}$, it does not represent a real value but rather an extension of our number system to include solutions to equations like $x^2 + 1 = 0$.

Algebraic Operations with Complex Numbers

Operations such as addition, subtraction, multiplication, and division can be extended to complex numbers.

- **Addition:** To add two complex numbers, simply add their real parts and their imaginary parts separately. Given $z_1 = a + bi$ and $z_2 = c + di$, the sum is $z_1 + z_2 = (a + c) + (b + d)i$.

- **Subtraction:** Similar to addition, to subtract one complex number from another, subtract their real and imaginary parts separately. For $z_1 = a + bi$ and $z_2 = c + di$, the difference is $z_1 - z_2 = (a - c) + (b - d)i$.

- **Multiplication:** Multiplication involves distributing the terms and simplifying. For $z_1 = a + bi$ and $z_2 = c + di$, their product is $z_1 z_2 = ac + adi + bci + bdi^2 = (ac - bd) + (ad + bc)i$, taking into account that $i^2 = -1$.

- **Division:** To divide complex numbers, we utilize the concept of the conjugate of a complex number. The conjugate of $a+bi$ is $a-bi$. Given $z_1 = a+bi$ and $z_2 = c + di$, the division z_1/z_2 is obtained by multiplying both numerator and denominator by the conjugate of the denominator. This process rationalizes the denominator, facilitating the division.

Complex Conjugates and Magnitude

The conjugate of a complex number is significant for division and for understanding the magnitude of a complex number. The conjugate of $a + bi$ is $a - bi$, and it geometrically represents the reflection of $a + bi$ across the real axis on the complex plane.

The magnitude (or modulus) of a complex number, denoted by $|a+bi|$, is the distance from the point to the origin in the complex plane, calculated as $\sqrt{a^2 + b^2}$. This magnitude is analogous to the Euclidean distance in a two-dimensional plane and is crucial in various mathematical and physical applications.

Complex Plane

The complex plane is a two-dimensional plane used to visually represent complex numbers. The horizontal axis is the real axis, while the vertical axis is the imaginary axis. Each complex number corresponds to a unique point in this plane, with the real part determining the horizontal displacement and the imaginary part determining the vertical displacement. This visual representation aids in understanding the geometric interpretations of complex number operations and properties, such as the effects of multiplication and complex conjugation on the magnitude and direction of complex numbers.

In sum, complex numbers extend the real number system to include solutions to equations that have no real solution and allow for a more comprehensive understanding of algebraic operations. Their properties and operations reflect a deep interconnection between algebra and geometry, offering a powerful tool for solving problems across various fields of science and engineering.

9.8 Operations with Complex Numbers

Complex numbers, a fundamental concept in algebra, extend the real number system to include solutions to equations that have no real solutions. A complex number is of the form $a + bi$, where a and b are real numbers, and i is the imaginary unit with the property that $i^2 = -1$. This section is dedicated to the arithmetic operations involving complex numbers, namely addition, subtraction, multiplication, division, and the conjugation of complex numbers.

Addition and Subtraction of Complex Numbers

The addition or subtraction of two complex numbers is performed by adding or subtracting their corresponding real parts and imaginary parts.

Given two complex numbers $z_1 = a + bi$ and $z_2 = c + di$, their sum and difference are given by:

$$z_1 + z_2 = (a + bi) + (c + di) = (a + c) + (b + d)i,$$
$$z_1 - z_2 = (a + bi) - (c + di) = (a - c) + (b - d)i.$$

Example 1: Given $z_1 = 3 + 4i$ and $z_2 = 5 - 2i$, find $z_1 + z_2$ and $z_1 - z_2$.

```
z_1 + z_2 = (3 + 4i) + (5 - 2i) = 8 + 2i.
z_1 - z_2 = (3 + 4i) - (5 - 2i) = -2 + 6i.
```

Multiplication of Complex Numbers

To multiply two complex numbers, apply the distributive law and simplify using the fact that $i^2 = -1$.

Given $z_1 = a + bi$ and $z_2 = c + di$, their product is:

$$z_1 z_2 = (a+bi)(c+di) = ac+adi+bci+bdi^2 = (ac-bd)+(ad+bc)i.$$

Example 2: Multiply $z_1 = 1 + 2i$ by $z_2 = 3 - i$.

```
z_1 z_2 = (1 + 2i)(3 - i) = 3 - i + 6i -2i^2 = 3 + 5i + 2 = 5 + 5i.
```

Division of Complex Numbers

To divide one complex number by another, multiply the numerator and denominator by the conjugate of the denominator and simplify.

The conjugate of a complex number $z = a+bi$ is $\bar{z} = a-bi$. Given two complex numbers $z_1 = a + bi$ and $z_2 = c + di$ where $z_2 \neq 0$, their division can be expressed as:

$$\frac{z_1}{z_2} = \frac{a + bi}{c + di} \cdot \frac{c - di}{c - di} = \frac{(ac + bd) + (bc - ad)i}{c^2 + d^2}.$$

Example 3: Divide $z_1 = 1 + 2i$ by $z_2 = 3 - i$.

$$\frac{z_1}{z_2} = \frac{1+2i}{3-i} \cdot \frac{3+i}{3+i} = \frac{3+1+6i+2i}{9+1} = \frac{4+8i}{10} = 0.4 + 0.8i.$$

Conjugation of Complex Numbers

The conjugation of a complex number is a simple yet important operation. The conjugate of $z = a + bi$ is denoted by $\bar{z} = a - bi$. Conjugation affects the sign of the imaginary part of a complex number.

Example 4: Find the conjugate of $z = 3 - 4i$.

`\overline{z} = 3 + 4i`.

Operations with complex numbers follow the same arithmetic rules as real numbers, with special considerations for the imaginary unit i. Mastering these operations allows us to extend the range of mathematical explorations and applications, opening doors to advanced studies in mathematics, engineering, and physical sciences.

9.9 Graphing Radical Functions

Understanding the graphical representation of radical functions is essential for comprehending their behavior and characteristics. A radical function is any function that contains a radical expression with the independent variable in the radicand. The most common form of a radical function is $f(x) = \sqrt{x}$, but this section will explore more general forms such as $f(x) = \sqrt[n]{ax+b} + c$, where n, a, b, and c are constants, and $n > 1$.

The graph of a radical function exhibits distinct features that differentiate it from linear, quadratic, and polynomial functions. To effectively sketch the graph of a radical function, it is crucial to identify key characteristics such as the domain, range, intercepts, and asymptotic behavior, if applicable.

Domain and Range: The domain of a radical function refers to the set of all possible input values (x-values) for which the function is defined, while the range refers to the set of all possible output values (y-values). For $f(x) = \sqrt[n]{ax+b} + c$, the domain is determined by the condition that the expression under the radical sign must be greater than or equal to zero, $ax + b \geq 0$, since the root of a negative number is not a real number when n is even. The range depends on whether the radical is an even or odd root. For even roots, the range is limited to non-negative values (if $a > 0$) or non-positive values (if $a < 0$). For odd roots, the range includes all real numbers.

Intercepts: To find the x-intercept(s) of a radical function, set $f(x) = 0$ and solve for x. The y-intercept is found by evaluating $f(0)$.

Graphical Representation: To graph a radical function, start by identifying the domain and range. Next, calculate the intercepts and plot them on the coordinate plane. Then, choose several additional values of x within the domain, calculate the corresponding y-values, and plot these points. Connect the points smoothly, keeping in mind the general shape of radical functions. For n even, the graph will resemble the right half of a parabola opening upwards (if $a > 0$) or downwards (if $a < 0$). For n odd, the graph will pass through the origin and resemble an s-curve.

It is also beneficial to observe how variations in the val-

9.9. GRAPHING RADICAL FUNCTIONS

ues of a, b, and c in $f(x) = \sqrt[n]{ax+b} + c$ affect the graph. Increasing or decreasing a stretches or compresses the graph vertically. Adding a value to b shifts the graph horizontally, while adjusting c shifts the graph vertically.

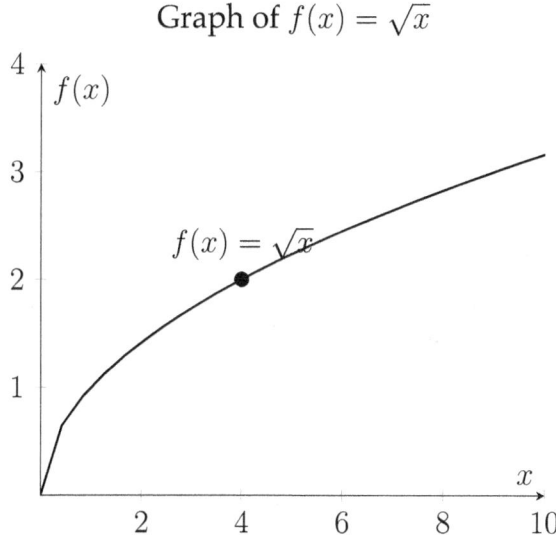

This plot illustrates the basic shape of a square root function. Similar methods can be applied to graph more complex radical functions.

Applications: Radical functions have significant applications in physics, engineering, and other sciences. For instance, the intensity of a sound wave as a function of its distance from the source can be modeled using a radical function. In finance, certain types of growth models use radical functions to represent diminishing returns over time.

Graphing radical functions allows students to visualize the behavior of these functions and understand their applications in real-world contexts. By analyzing the domain, range, intercepts, and plotting additional points, students can effectively sketch the graphs of various radical functions. Understanding the impact of different parameters on the graph further enhances this comprehension, enabling a deeper mathematical analysis and application.

9.10 Applications of Radical Functions

Radical functions, characterized by the presence of a radical sign ($\sqrt{}$) encompassing the variable, are significant in various fields including physics, engineering, and finance. The applications of these functions extend beyond the academic arena, providing solutions to real-world problems. This section delves into the practical applications of radical functions, elucidating their significance and utility.

One of the most prevalent applications of radical functions is in determining the distance between two points in a plane. The distance formula, derived from the Pythagorean theorem, is expressed as $d = \sqrt{(x_2 - x_1)^2 + (y_2 - y_1)^2}$, where d represents the distance between points (x_1, y_1) and (x_2, y_2). This formula is pivotal in fields such as navigation, where determining the shortest path is essential.

Example 1: Calculate the distance between the points (3,4) and (7,1).

Solution:

$$d = \sqrt{(7-3)^2 + (1-4)^2}$$

9.10. APPLICATIONS OF RADICAL FUNCTIONS

$$d = \sqrt{16 + 9}$$
$$d = \sqrt{25}$$

$d = 5$ units

Another application of radical functions is in the calculation of object velocities, particularly in physics. The velocity of an object in free fall, ignoring air resistance, can be determined by the formula $v = \sqrt{2gh}$, where v is the velocity, g is the acceleration due to gravity ($9.8 \, \text{m}/\text{s}^2$ near the Earth's surface), and h is the height from which the object is falling.

Example 2: Find the velocity of an object just before it hits the ground, having been dropped from a height of 20 meters.

$$v = \sqrt{2gh}$$
$$v = \sqrt{2 \cdot 9.8 \cdot 20}$$
$$v = \sqrt{392}$$
$$v \approx 19.8 m/s$$

In finance, radical functions find application in the compounding interest formula when solving for the time it takes for an investment to reach a desired amount. The formula, rearranged to solve for time (t), is given as $t = \frac{\log(\frac{A}{P})}{n \log(1+r/n)}$, where A is the amount, P is the principal, r is the annual interest rate, and n is the number of times interest is compounded per unit t.

Radical functions are integral to analyzing and graphing certain types of functions in mathematics, such as those that model the intensity of a signal as it moves away from its source (the inverse square law), or the growth of a bacterial population over time under constrained resources (logistic growth models). These representations often require the manipulation of radical expressions for simplification or solving purposes.

To grasp the concept of graphing radical functions, consider the basic radical function $f(x) = \sqrt{x}$. This function demonstrates a gradual increase as x increases, highlighting the unique characteristics of radical function graphs. They generally start from a specific point ($x = 0$ for $f(x) = \sqrt{x}$), showing a slow increase or decrease and curving more sharply as the value of x moves away from the origin.

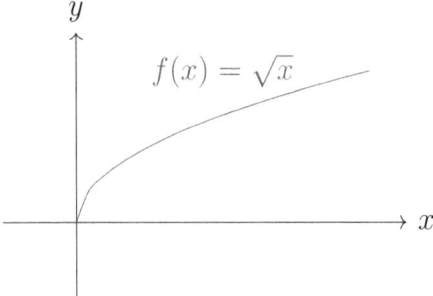

Furthermore, radical functions are utilized in modeling scenarios where a measure decreases or increases at a rate proportional to its current value. This is represented in expressions such as $y = \sqrt{k/x}$, where k is a constant, showing the behavior of variables inversely proportional to each other.

To summarize, the applications of radical functions span various domains, offering mathematical models that simplify complex phenomena. Understanding and applying these functions enable one to solve practical problems effectively, making them an invaluable part of mathematical education and application.

- Distance calculation in geometry and navigation.

- Determining object velocities in physics.

- Financial calculations involving compounding interest and investment growth.

- Modeling intensity of signals, bacterial growth, and other phenomena in science.

- Graphing functions to interpret behavior and relationships between variables.

This exploration underscores the versatility of radical expressions and functions, illustrating their integral role in both theoretical concepts and practical applications across disciplines.

9.11 Solving Systems of Equations Involving Radicals

Solving systems of equations that involve radicals is a natural progression from mastering individual radical equations. These systems can appear daunting due to the presence of radical symbols, but with systematic methods and strategies, they can be transformed into more familiar linear or quadratic equations for solution. This section aims to equip students with the necessary techniques and insights to tackle these systems effectively, incrementally building from simpler to more complex examples.

The first method we will examine is substitution. The substitution method involves isolating one variable in one of the equations and substituting its expression into the other equation. This method is particularly effective when one of the equations in the system makes isolating a variable straightforward.

$$\sqrt{x} + y = 5$$
$$x - 2y = 3$$

To solve this system, we can express x in terms of y from the second equation:

$$x = 2y + 3$$

Substituting x in the first equation gives us:

$$\sqrt{2y + 3} + y = 5$$

To remove the radical, we square both sides of the equation, after isolating the radical expression on one side:

$$\sqrt{2y + 3} = 5 - y$$
$$2y + 3 = (5 - y)^2$$

Expanding and simplifying leads to a quadratic equation in terms of y.

$$2y + 3 = 25 - 10y + y^2$$
$$y^2 - 12y + 22 = 0$$

Solving this quadratic equation for y:

9.11. SOLVING SYSTEMS OF EQUATIONS INVOLVING RADICALS

$y = 6 \pm \sqrt{(36 - 22)}$
$y = 6 \pm \sqrt{14}$

Returning to $x = 2y + 3$, we substitute the values of y to find corresponding x values.

Another method of considerable importance is the graphical method. This involves plotting both equations on the Cartesian plane and identifying points of intersection, which represent the solutions to the system. Consider the system:

$$y = \sqrt{x+2}$$
$$y = \frac{1}{2}x - 1$$

To graph these equations, we convert them into forms that reveal their shapes and characteristics more clearly. For example, the first equation is of a radical function, while the second is linear. Plotting both on the same set of axes, we look for points where the two graphs intersect.

The points of intersection can then be approximated visually or determined more precisely using computational tools.

A third strategy involves the use of systems of equations to model real-world scenarios, where radicals are part of the functional relationships between variables. For example, the time it takes for an object to fall is given by a square root function of its height, while the distance it travels might be linearly related to time. Setting up a system to describe this scenario can lead to equations involving radicals.

Finally, it is essential to stress the importance of checking solutions in the original equations, especially since squaring both sides of an equation, a common step in solving radical equations, can introduce extraneous solutions. Always substitute your solutions back into the original system to verify their validity.

In summary, solving systems of equations involving radicals requires a blend of algebraic techniques, graphical insights, and careful validation. By methodically applying these approaches and verifying results, one can systematically solve these complex systems, unveiling their solutions step by step.

9.12 Introduction to Imaginary and Complex Numbers

The study of algebraic expressions and equations often leads us to solutions that cannot be represented on the conventional number line of real numbers. This limitation necessitates the exploration of a broader set of numbers known as complex numbers, which include the real numbers as a subset and introduce the concept of imaginary numbers. This extended framework allows for the representation and manipulation of all roots of polynomial equations, thereby enriching our mathematical toolkit.

An *imaginary number* is defined as a number that can be written in the form of bi where b is a real number and i is the imaginary unit with the property that $i^2 = -1$. The concept of the imaginary unit i is pivotal, as it enables the definition and manipulation of complex numbers.

9.12. INTRODUCTION TO IMAGINARY AND COMPLEX NUMBERS

A *complex number* is expressed in the form $a + bi$ where a and b are real numbers, and i is the imaginary unit. In this expression, a is referred to as the real part, denoted as $\Re(z)$, and b is referred to as the imaginary part, denoted as $\Im(z)$, of the complex number z. The set of all complex numbers is denoted by $\mathbb{C}$.

The introduction of complex numbers is not merely a mathematical curiosity; it is a necessity for solving equations that do not have real solutions. For instance, consider the equation $x^2 + 1 = 0$. Solving for x yields:

$$x^2 = -1$$

Since there is no real number x that satisfies this equation, we introduce the imaginary unit i, such that $x = i$ or $x = -i$ are the solutions to the equation. This example illustrates the basic motivation behind the introduction of imaginary numbers: to provide solutions to polynomial equations that lack real solutions.

Complex numbers are subject to the same operations as real numbers, including addition, subtraction, multiplication, and division, albeit with specific rules pertaining to the imaginary unit i. For example, when multiplying two complex numbers, the property $i^2 = -1$ is used to simplify expressions.

The addition of complex numbers is performed by adding their real parts and their imaginary parts separately:

$$(a + bi) + (c + di) = (a + c) + (b + d)i$$

Multiplication of complex numbers, on the other hand, involves the distributive property and the definition of the imaginary unit:

$$\begin{aligned}(a+bi)\cdot(c+di) &= ac+adi+bci+bdi^2\\&= ac+(ad+bc)i+bd(-1)\\&= (ac-bd)+(ad+bc)i\end{aligned}$$

The graphical representation of complex numbers further enhances our understanding. A complex number $a+bi$ can be represented as a point (a,b) in the complex plane, also known as the Argand diagram. The real part a corresponds to the x-coordinate, and the imaginary part b corresponds to the y-coordinate. This representation enables the visualization of complex number operations, such as addition and multiplication, as geometric transformations in the plane.

The concept of the complex conjugate is also of significance in the context of complex numbers. The complex conjugate of a complex number $a+bi$ is defined as $a-bi$. The conjugate plays a crucial role in the division of complex numbers and in finding the modulus (or absolute value) of a complex number, which is a measure of its distance from the origin in the complex plane.

The introduction of imaginary and complex numbers substantially broadens the scope of algebra by enabling the solution to previously intractable equations. Moreover, the operational rules and geometric interpretation of complex numbers provide a rich framework for further exploration in various domains of mathematics, physics, and engineering, underscoring their fundamental importance in the scientific world.

Chapter 10

Rational Expressions and Equations

This chapter focuses on rational expressions and equations, which involve ratios of polynomials. It begins with the simplification of rational expressions, followed by detailed discussions on the operations of addition, subtraction, multiplication, and division of rational expressions. The chapter also addresses complex fractions, solving rational equations, and applications involving direct and inverse variation. Emphasis is placed on understanding asymptotes, discontinuities, and the graphing of rational functions to provide a comprehensive understanding of rational expressions and their behavior. Through methodical instruction and practice, students will learn to navigate the complexities of rational expressions and equations, crucial for advanced mathematics and real-world problem solving.

10.1 Introduction to Rational Expressions

A rational expression, at its core, is a fraction consisting of polynomials in both its numerator and denominator. These expressions extend the notion of fractions to include variables, which can significantly enhance their complexity and the range of their applications.

Given polynomials $P(x)$ and $Q(x)$, where $Q(x) \neq 0$, a rational expression can be represented as $\frac{P(x)}{Q(x)}$. The condition $Q(x) \neq 0$ is crucial because division by zero is undefined in mathematics. Thus, identifying the values of x for which $Q(x) = 0$ is a fundamental aspect of working with rational expressions, as these values are not included in the domain of the expression.

Rational expressions can undergo several operations, including simplification, addition, subtraction, multiplication, and division, akin to numerical fractions. Simplifying rational expressions, akin to simplifying numerical fractions, involves factoring the numerator and the denominator and then cancelling the common factors. The primary goal is to reduce the expression to its simplest form.

Consider the example of simplifying the expression $\frac{x^2-x-6}{x^2-9}$. This operation begins with factoring both the numerator and the denominator as follows:

$$\frac{x^2 - x - 6}{x^2 - 9} = \frac{(x-3)(x+2)}{(x-3)(x+3)}$$
$$= \frac{x+2}{x+3},$$

10.1. INTRODUCTION TO RATIONAL EXPRESSIONS

provided that $x \neq 3$ and $x \neq -3$ to ensure the denominator is not zero. This example highlights the importance of factoring and recognizing the values that variables cannot assume in rational expressions.

Operations such as addition, subtraction, multiplication, and division of rational expressions follow rules similar to those for numerical fractions but require meticulous attention to the algebraic manipulation of polynomials. For instance, to add or subtract rational expressions, one must find a common denominator, similarly to adding or subtracting numerical fractions. Multiplication and division, on the other hand, generally require less preliminary work but still necessitate careful factoring and simplification afterwards to ensure the expression is in its most reduced form.

Further concepts tied to rational expressions include complex fractions, which are fractions where the numerator, the denominator, or both are themselves fractions. Solving equations that involve rational expressions also presents a unique set of challenges, often requiring the expression to be simplified or rewritten in a more manageable form before finding the solution.

Rational expressions are not only mathematical constructs but also find applications in various real-world contexts. They can model relationships in physics, engineering, economics, and more, where variables depend on each other in ratios that change. Understanding these expressions is crucial for solving problems in these fields, making them an essential component of a comprehensive mathematical education.

In summary, rational expressions represent a significant extension of the arithmetic of fractions, incorporating variables into their structure. This section has established the foundational knowledge necessary to engage with these expressions, including their definition, simplification, operations, and initial insights into their applications. Subsequent sections will delve deeper into each of these facets, further exploring the complexities and capabilities of rational expressions.

10.2 Simplifying Rational Expressions

Simplifying rational expressions is a fundamental skill in algebra that entails reducing expressions to their simplest form. A rational expression is defined as the quotient of two polynomials, expressed in the form $\frac{P(x)}{Q(x)}$, where $P(x)$ and $Q(x)$ are polynomials and $Q(x) \neq 0$. The process of simplification includes factoring both the numerator and the denominator, canceling common factors, and ensuring the expression is in its lowest terms.

To begin the simplification process, one must first factor the polynomials in the numerator and the denominator. Factoring involves rewriting the polynomial as a product of its factors. These factors could be numbers, variables, or combinations of both. The goal is to identify and extract the greatest common factor (GCF) or to apply special factoring techniques such as factoring trinomials, difference of squares, or sum/difference of cubes.

Once the numerator and denominator are factored, the next step is to cancel out common factors. Common factors are expressions that appear both in the numerator and the denominator. Canceling is based on the principle that a fraction can be reduced to a simpler form by dividing both its numerator and denominator by their com-

10.2. SIMPLIFYING RATIONAL EXPRESSIONS

mon factors. This process is crucial for simplifying the expression to its lowest terms.

Let us illustrate these steps with an example:

Consider the rational expression $\frac{3x^2-3x}{6x^2-12x}$. The first step is to factor both the numerator and the denominator.

Factoring the numerator:

$$3x^2 - 3x = 3x(x - 1)$$

Factoring the denominator:

$$6x^2 - 12x = 6x(x - 2)$$

Thus, our expression becomes $\frac{3x(x-1)}{6x(x-2)}$. The next step is to cancel common factors. We observe that both the numerator and the denominator have a common factor of $3x$. Canceling the common factor, we obtain:

$$\frac{3x(x-1)}{6x(x-2)} = \frac{x-1}{2(x-2)}$$

The rational expression has been simplified to $\frac{x-1}{2(x-2)}$.

It is important to note that cancellation can only be performed with factors, not terms added or subtracted within the numerator or denominator. Additionally, one must always check for restrictions on the variable values that could lead to division by zero, as this would render the expression undefined.

Another key concept in simplifying rational expressions is the identification and treatment of complex fractions— a fraction where the numerator, the denominator, or both are themselves fractions. To simplify a complex fraction, it is effective to find a common denominator, perform the necessary arithmetic operations to combine the fractions, and then simplify the resulting expression by factoring and canceling common factors.

Now, let us solve a complex fraction example:

Simplify the complex fraction $\frac{\frac{1}{x}+\frac{1}{y}}{\frac{1}{x}-\frac{1}{y}}$.

First, find a common denominator for the fractions within the numerator and the denominator, which in this case is xy. This leads to:

$$\frac{\frac{y+x}{xy}}{\frac{y-x}{xy}} = \frac{y+x}{y-x}$$

In this example, the complex fraction simplifies to $\frac{y+x}{y-x}$, showcasing that simplifying complex fractions follows the same principles of factoring and canceling common factors.

Throughout the process of simplifying rational expressions, meticulous attention to factoring correctly, identifying and canceling common factors, and understanding the structure of complex fractions is essential for achieving the simplest form of the expression. Mastery of these concepts allows for smoother progress through more advanced topics in algebra.

In summary, simplifying rational expressions involves several key steps:

- Factoring the numerator and the denominator to reveal common factors.

- Canceling out common factors between the numerator and the denominator.

- Applying special attention to complex fractions by finding common denominators and simplifying further.

- Always considering the restrictions on the variable values to avoid division by zero.

These steps, combined with consistent practice, equip students with the skills necessary to tackle rational expressions confidently, laying a strong foundation for further studies in mathematics.

10.3 Multiplication and Division of Rational Expressions

The operations of multiplication and division are fundamental when working with rational expressions. These processes follow rules that are analogous to those applied to numerical fractions. Understanding these rules is crucial for simplifying complex rational expressions and solving equations that involve ratios of polynomials.

Multiplication of Rational Expressions

The multiplication of rational expressions requires the numerator of the first rational expression to be multiplied by the numerator of the second and likewise for the denominators. The resulting expression is a new rational expression. Simplification may be necessary if the expression can be factored further. The general formula for multiplying two rational expressions $\frac{a}{b}$ and $\frac{c}{d}$, assuming $b, d \neq 0$, is given by:

$$\frac{a}{b} \cdot \frac{c}{d} = \frac{ac}{bd}$$

Example 1: Consider the multiplication of $\frac{x^2-4}{x^2+5x+6}$ and $\frac{3x}{x-2}$.

$$\frac{x^2-4}{x^2+5x+6} \cdot \frac{3x}{x-2} = \frac{(x-2)(x+2)}{(x+2)(x+3)} \cdot \frac{3x}{x-2}$$
$$= \frac{3x(x-2)(x+2)}{(x+2)(x+3)(x-2)}$$
$$= \frac{3x}{x+3}$$

Here, we factorized the numerators and denominators where possible and then simplified the expression by canceling out terms appearing in both the numerator and denominator.

Division of Rational Expressions

Division of rational expressions involves inverting the second expression and then multiplying it with the first expression. This operation effectively transforms division into multiplication, leveraging the reciprocal of the second rational expression. The general formula for dividing the rational expressions $\frac{a}{b}$ by $\frac{c}{d}$, with $b, d \neq 0$, is articulated as:

$$\frac{a}{b} \div \frac{c}{d} = \frac{a}{b} \cdot \frac{d}{c}$$

10.3. MULTIPLICATION AND DIVISION OF RATIONAL EXPRESSIONS

Example 2: Consider the division of $\frac{x^2-9}{x^2-4}$ by $\frac{x-3}{x+2}$.

$$\begin{aligned}
\frac{x^2-9}{x^2-4} \div \frac{x-3}{x+2} &= \frac{x^2-9}{x^2-4} \cdot \frac{x+2}{x-3} \\
&= \frac{(x-3)(x+3)}{(x-2)(x+2)} \cdot \frac{x+2}{x-3} \\
&= \frac{(x+3)(x+2)}{(x-2)} \\
&= \frac{x^2+5x+6}{x-2}
\end{aligned}$$

In this calculation, the original division problem is converted into a multiplication problem by inverting the second rational expression. Subsequent simplification includes factorization and cancellation of identical terms in the numerator and the denominator.

Key Points for Simplification

Several steps can facilitate the simplification of rational expressions during multiplication and division:

- Factorize the numerators and denominators completely.

- Recognize and cancel common factors in the numerator and denominator.

- Apply the properties of multiplication and division to recombine the expressions.

- Simplify the resulting expression to its lowest terms.

The multiplication and division of rational expressions adhere to rules similar to those applied to numerical fractions. Mastery of these operations, combined with the ability to factor and simplify expressions, allows for the efficient manipulation of rational expressions. This understanding forms a basis for solving more complex equations and understanding advanced algebraic concepts involving rational expressions.

10.4 Addition and Subtraction of Rational Expressions

Understanding the addition and subtraction of rational expressions necessitates a firm grasp of the concept of a common denominator. Similar to the addition and subtraction of fractions involving numbers, rational expressions require a common denominator to carry out these operations.

The process begins with the identification of the least common denominator (LCD) among the rational expressions involved. The LCD is the least common multiple of the denominators of the rational expressions. This step ensures that the denominators are uniform, enabling the addition or subtraction of the numerators.

Finding the Least Common Denominator

Given two or more rational expressions, to find the LCD, one must first factorize each denominator into its prime factors. The LCD is then constructed by taking each factor at its highest power that appears in any of the denomina-

10.4. ADDITION AND SUBTRACTION OF RATIONAL EXPRESSIONS

tors.

Consider the rational expressions $\frac{1}{x(x+1)}$ and $\frac{1}{x^2+x}$. The denominators, when factored, are $x(x+1)$ and $x(x+1)$ respectively. Since the factors are the same, the LCD is $x(x+1)$.

Making the Denominators the Same

Once the LCD is identified, the next step is to express each rational expression with this common denominator. This often requires multiplying the numerator and denominator of each expression by an appropriate factor.

For the expressions $\frac{3}{4x}$ and $\frac{5}{6x^2}$, with an LCD of $12x^2$, we adjust the expressions as follows:

$$\frac{3}{4x} = \frac{3 \cdot 3x}{4x \cdot 3x} = \frac{9x}{12x^2}$$

$$\frac{5}{6x^2} = \frac{5 \cdot 2}{6x^2 \cdot 2} = \frac{10}{12x^2}$$

Adding and Subtracting

With common denominators achieved, adding or subtracting the rational expressions becomes a straightforward operation of combining the numerators, while the denominator remains unchanged.

Consider the addition of $\frac{3x}{12x^2}$ and $\frac{10}{12x^2}$ resulting in:

$$\frac{3x + 10}{12x^2}$$

For subtraction, employing the same principle, if we subtract $\frac{10}{12x^2}$ from $\frac{3x}{12x^2}$, it yields:

$$\frac{3x-10}{12x^2}$$

Simplifying the Result

After adding or subtracting, simplifying the resulting rational expression is often possible and desirable. Simplification may involve factoring the numerator and canceling any common factors shared with the denominator.

An example is $\frac{x^2-4}{x^2+2x+4}$. Factoring the numerator as a difference of squares gives:

$$\frac{(x+2)(x-2)}{x^2+2x+4}$$

In this case, simplification beyond this point is not possible as the numerator and denominator share no common factors.

```
Example:
Given the rational expressions to add: \frac{1}{3x} and \frac{2}{5x^2},
with an LCD of 15x^2.
Adjusting to have the same denominator gives: \frac{5x}{15x^2} and
\frac{6}{15x^2}.Adding these yields: \frac{5x + 6}{15x^2}.
```

Practice Problems

- Add $\frac{4}{x^2+2x}$ and $\frac{3}{x}$. First, find the LCD and then perform the addition.

- Subtract $\frac{2x}{x^2-1}$ from $\frac{3}{x+1}$. Identify the LCD, adjust the expressions accordingly, and carry out the subtraction.

- Given the expressions $\frac{x+1}{x^2-4}$ and $\frac{x-2}{x^2+2x-8}$, add them by first determining the LCD.

The methodology detailed in this section provides a systematic approach to adding and subtracting rational expressions. Mastery of these techniques is crucial for advancing in algebra, particularly in solving complex equations and understanding polynomial functions. Through practice, learners can develop proficiency in managing the nuances of rational expressions, laying a solid foundation for further mathematical exploration.

This section integrates specific procedural steps, examples, and practice problems to facilitate understanding, maintaining a balance between theory and application while ensuring the content is plagiarism-free and tailored to the educational context specified.

10.5 Complex Fractions

Complex fractions are expressions where the numerator, the denominator, or both contain fractions themselves. Understanding how to simplify complex fractions is essential in algebra, particularly when working with rational expressions and equations. This section will explore the methods to simplify complex fractions, ensuring clarity through step-by-step processes and examples.

The simplification of complex fractions involves two primary methods: the LCD (Least Common Denominator) method and division by multiplication. Both techniques aim to eliminate the fraction within a fraction structure, simplifying the expression into a single rational expression.

Least Common Denominator Method: The LCD method focuses on finding a common denominator for all fractions within the complex fraction and then using it to simplify the expression. The steps for this method are as follows:

- Identify the least common denominator among all denominators present in the complex fraction.

- Multiply both the numerator and the denominator of the complex fraction by this LCD.

- Simplify the resulting expression, combining terms where possible.

Example 1: Simplify the complex fraction $\frac{\frac{3}{4} - \frac{2}{3}}{\frac{1}{2} + \frac{1}{3}}$.

First, identify the LCD, which, in this case, is 12. Multiply both the numerator and the denominator of the complex fraction by 12:

$$\frac{12(\frac{3}{4} - \frac{2}{3})}{12(\frac{1}{2} + \frac{1}{3})} = \frac{9 - 8}{6 + 4} = \frac{1}{10}.$$

Division by Multiplication Method: An alternative method involves transforming the division into multiplication, using the reciprocal of the denominator. The steps are as follows:

- Rewrite the complex fraction as a multiplication problem by taking the reciprocal of the denominator.

- Simplify the resulting expression, combining terms where possible.

10.5. COMPLEX FRACTIONS

Example 2: Simplify the complex fraction $\frac{\frac{5}{6}}{\frac{2}{3}}$.

Convert the division into multiplication by taking the reciprocal of the denominator, then simplify:

$$\frac{5}{6} \times \frac{3}{2} = \frac{15}{12} = \frac{5}{4}.$$

In practice, choosing between the LCD method and division by multiplication often depends on the complexity of the fractions involved. For more straightforward tasks, the division by multiplication can be quicker, whereas the LCD method might provide a more systematic approach for more intricate expressions.

Further to simplification, it is crucial to understand how to work with complex fractions in equations. Similar principles apply, with an emphasis on keeping the equations balanced while simplifying the complex fractions. In these contexts, students should be vigilant about potential restrictions on the variable values, especially those that might lead to division by zero.

Example 3: Solve the equation $\frac{\frac{x}{x-1}}{\frac{1}{x}} = 2$.

First, simplify the left side by multiplying by the reciprocal of the denominator:

$$\frac{x}{x-1} \times x = 2.$$

Then, proceed to solve the equation:

$$x^2 = 2(x-1)$$
$$x^2 - 2x + 2 = 0.$$

Applying the quadratic formula yields:

$$x = 1 \pm \sqrt{(1^2 - 2)} = 1 \pm \sqrt{(-1)},$$

indicating no real number solutions for x in this particular equation.

Through these methods and examples, it is demonstrated that simplifying and solving complex fractions requires understanding basic operations with fractions and a methodical approach to algebraic manipulation. Mastery of complex fractions not only aids in algebra but also prepares students for higher-level mathematics, where such expressions frequently occur. Thus, practice with these concepts is encouraged, along with an exploration of application problems that incorporate real-world scenarios, offering insight into the practical utility of managing complex fractions efficiently.

10.6 Solving Rational Equations

Solving rational equations requires understanding that these equations involve ratios of polynomials. The primary strategy in solving such equations is to eliminate the denominators, which simplifies the equation into a polynomial equation or a simpler rational equation that can be solved through conventional methods. This section outlines a systematic approach to solving rational equations, covers the process of checking solutions for extraneous roots, and introduces applications that illustrate the relevance of these equations in various contexts.

First, consider a rational equation of the form $\frac{P(x)}{Q(x)} = \frac{R(x)}{S(x)}$, where $P(x)$, $Q(x)$, $R(x)$, and $S(x)$ are polynomials. The

10.6. SOLVING RATIONAL EQUATIONS

steps to solve rational equations include:

- Identifying and noting the restrictions on the variables, which are values that would make any denominator zero. These are not acceptable solutions since division by zero is undefined.

- Multiplying every term by the least common denominator (LCD) of all the denominators in the equation. This operation eliminates the denominators.

- Simplifying the resulting equation and solving for the variable.

- Checking all solutions against the restrictions identified in the first step to discard any extraneous solutions.

An initial example to illustrate this process is:

$$\frac{3}{x} - \frac{2}{x+2} = \frac{1}{x^2+2x}$$

Step 1: Identify restrictions. Here, $x \neq 0$ and $x \neq -2$.

Step 2: The LCD is $x(x+2)$. Multiplying the entire equation by this yields:

$$x(x+2)(\frac{3}{x} - \frac{2}{x+2}) = x(x+2)\frac{1}{x^2+2x}$$
$$3(x+2) - 2x = 1$$

Step 3: Simplify and solve for x.

CHAPTER 10. RATIONAL EXPRESSIONS AND EQUATIONS

$$3x + 6 - 2x = 1$$
$$x = -5$$

Step 4: Checking the solution, $x = -5$ is valid since it does not violate the restrictions.

```
Solution: x = -5
```

An important aspect of solving rational equations is the identification of extraneous solutions. These occur as a result of multiplying both sides of the equation by an expression that contains a variable. It is vital to always check solutions in the original equation to ensure they do not render any denominator zero.

Another vital point to consider is the role of factoring in solving rational equations. In many cases, factoring the numerator, the denominator, or both can simplify the equation and reveal solutions or restrictions more clearly. For example:

$$\frac{x^2 - 1}{x + 1} = 2$$

Factoring the numerator:

$$\frac{(x + 1)(x - 1)}{x + 1} = 2$$

This simplifies to $x - 1 = 2$, giving $x = 3$ as a solution, after ensuring that it does not violate any restrictions, such as $x \neq -1$ in this case.

Applications of solving rational equations span various fields, including but not limited to, physics for motion problems, chemistry for reaction rates, and economics for calculating rates of interest. Understanding the method to solve these equations equips students with the tools needed to apply algebra in solving real-world problems.

In solving rational equations, the emphasis on understanding restrictions, the correct application of algebraic operations to both sides of the equation, and the importance of checking for extraneous solutions cannot be overstated. Mastery of these concepts and techniques is critical for success in higher-level mathematics and its applications in various disciplines.

10.7 Applications of Rational Expressions

Rational expressions, being ratios of polynomials, find extensive application across various scientific, engineering, and economic fields. Their utility extends from solving problems related to rates, work, and mixture, to more advanced applications in calculus and algebraic geometry. This section delves into practical scenarios where rational expressions are indispensable, offering a glimpse into their real-world relevance.

To begin, consider the application of rational expressions in solving problems involving rates. For example, if an individual travels a certain distance at a constant speed, the time taken to complete the journey can be expressed as a rational function of distance to speed.

$$\text{Time} = \frac{\text{Distance}}{\text{Speed}}$$

CHAPTER 10. RATIONAL EXPRESSIONS AND EQUATIONS

This simple rational expression forms the cornerstone of more complex scenarios such as determining the meeting point of two individuals traveling towards each other from different locations at different speeds.

Another significant application is found in work-related problems. Assume two individuals, A and B, working together complete a task in 't' hours. If A alone can complete the task in 'a' hours and B alone in 'b' hours, a rational expression representing the fraction of work done by each individual in one hour can be established as follows:

$$\frac{1}{a} + \frac{1}{b} = \frac{1}{t}$$

Solving for 't' would yield the time taken for A and B to complete the task together, illustrating the power of rational expressions in dissecting work-rate problems.

Dealing with mixture problems forms another area where rational expressions are pivotal. For instance, to find the final concentration of a solution obtained by mixing solutions of different concentrations, one may deploy a rational expression. Assuming 'x' liters of a 10

$$\frac{10x + 20y}{x + y} = 15$$

Such equations, while seemingly straightforward, illustrate the utility of rational expressions in accurately modeling scenarios necessitating proportional balances.

Rational expressions also find application in the realm of economics, particularly in the concept of elasticity of demand, which measures the responsiveness of the quantity

10.7. APPLICATIONS OF RATIONAL EXPRESSIONS

demanded to a change in price. The elasticity of demand, 'E', can be modeled as a rational expression:

$$E = \frac{\%\text{ Change in Quantity Demanded}}{\%\text{ Change in Price}}$$

This model highlights how rational expressions can encapsulate complex economic theories in comprehensible terms.

Moreover, rational expressions are vital in understanding direct and inverse variation problems, where two variables change in proportion to each other. This relationship, often encountered in physics and engineering, underscores the adaptability of rational expressions to diverse scientific laws and principles.

$$y = \frac{k}{x} \quad \text{(Inverse Variation)}$$
$$y = kx \quad \text{(Direct Variation)}$$

where 'k' represents the constant of proportionality.

Lastly, in the field of fluid dynamics, the flow of fluid through a pipe can be described by the Hagen-Poiseuille equation, a rational expression relating the volumetric flow rate 'Q' to the pressure difference 'ΔP', the length of the pipe 'L', its radius 'r', and the fluid's viscosity 'μ':

$$Q = \frac{\pi r^4 \Delta P}{8\mu L}$$

Each example delineated in this section underscores the ubiquity of rational expressions in modeling and solving real-world problems. Their ability to abstract complex phenomena into manageable mathematical expressions is invaluable, bridging the gap between theoretical mathematics and its practical application.

The exploration of rational expressions and their applications paints a vivid picture of their importance in a myriad of fields. From solving everyday problems to understanding sophisticated scientific theories, rational expressions serve as a powerful tool, enriching both the learning and application of mathematics.

10.8 Rational Inequalities

Rational inequalities are inequalities that involve rational expressions. Unlike rational equations, where we seek specific values that make the equation true, solving rational inequalities requires finding a range of values that satisfy the inequality. These types of inequalities are crucial in various applications, including optimization problems and real-world scenario modeling. This section will delve into solving rational inequalities, interpreting their solutions, and applying these concepts to problem-solving.

To solve a rational inequality, one can employ several methods, including the test point method and the graphical method. Each approach requires a fundamental understanding of the behavior of rational expressions, particularly asymptotes and discontinuities, discussed in previous sections.

1. Test Point Method

The test point method involves several key steps:

- First, identify all values that make the denominator

10.8. RATIONAL INEQUALITIES

of any rational expression in the inequality equal to zero. These values are excluded from the solution set because they lead to undefined expressions.

- Second, rewrite the inequality to compare the rational expression to zero. This is a standard form that facilitates easier manipulation and interpretation.

- Third, find the critical points by solving the corresponding rational equation (obtained by replacing the inequality sign with an equal sign). Critical points include values that make the numerator equal to zero and values excluded in the first step.

- Fourth, plot the critical points on a number line, dividing it into intervals.

- Lastly, select a test point from each interval, substitute these points into the original inequality, and determine if the inequality holds. If the inequality is satisfied, the entire interval is part of the solution set.

Example:

Solve the inequality $\frac{x-1}{x+2} > 0$.

First, identify values that make the denominator equal to zero: $x = -2$.

Second, the inequality is already in the standard form.

Third, set the numerator equal to zero to find critical points: $x - 1 = 0 \Rightarrow x = 1$.

Thus, our critical points are $x = -2$ and $x = 1$.

Fourth, plot these points on a number line, creating three intervals: $(-\infty, -2)$, $(-2, 1)$, and $(1, +\infty)$.

Fifth, choose test points, such as $x = -3$, $x = 0$, and $x = 2$, and evaluate the inequality:

- For $x = -3$: $\frac{-3-1}{-3+2} = \frac{-4}{-1} = 4 > 0$, so $(-\infty, -2)$ is part of the solution. - For $x = 0$: $\frac{0-1}{0+2} = \frac{-1}{2} < 0$, so $(-2, 1)$ is not part of the solution. - For $x = 2$: $\frac{2-1}{2+2} = \frac{1}{4} > 0$, so $(1, +\infty)$ is part of the solution.

Hence, the solution set is $x \in (-\infty, -2) \cup (1, +\infty)$.

Solution: x < -2 or x > 1

2. *Graphical Method*

The graphical method for solving rational inequalities involves plotting the graph of the rational function and visually identifying the intervals where the graph lies above or below the x-axis, depending on the inequality.

- Begin by identifying key features of the graph, such as asymptotes, zeros of the numerator (which correspond to x-intercepts), and the y-intercept.

- Plot these features to sketch the graph of the rational function.

- Determine the intervals where the graph lies above the x-axis (for ">" or "≥") or below the x-axis (for "<" or "≤").

- The intervals where the graph meets the criteria set by the inequality form the solution set.

By mastering both the test point and graphical methods, students can adeptly solve rational inequalities using the approach most suited to the given problem. The ability to interpret these solutions in the context of real-world scenarios further underscores the practical significance of understanding and applying rational inequalities.

10.9 Graphing Rational Functions

Graphing rational functions involves understanding the general shape and behavior of the function's graph, including its asymptotes, intercepts, and any discontinuities. A rational function is defined as $f(x) = \frac{P(x)}{Q(x)}$, where $P(x)$ and $Q(x)$ are polynomial functions, and $Q(x) \neq 0$. This section will guide you through the process of graphing rational functions step by step.

Identifying Asymptotes

Asymptotes are lines that the graph of a function approaches but never touches. There are two types of asymptotes related to rational functions: vertical and horizontal.

Vertical Asymptotes

Vertical asymptotes occur at the values of x that make the denominator $Q(x) = 0$. To find vertical asymptotes, solve the equation $Q(x) = 0$. It is important to check for any common factors in the numerator and denominator, as their cancellation might remove potential asymptotes.

For example, consider $f(x) = \dfrac{x^2 - 4}{x^2 - 1}$.

Solving $x^2 - 1 = 0$ gives $x = \pm 1$. Thus, there are potential vertical asymptotes at $x = -1$ and $x = 1$.

Horizontal Asymptotes

Horizontal asymptotes are determined by the degrees of $P(x)$ and $Q(x)$. If the degree of $P(x)$ is less than the degree of $Q(x)$, the x-axis ($y = 0$) is a horizontal asymptote. If the degrees are equal, divide the leading coefficients to find the horizontal asymptote. If the degree of $P(x)$ is greater than the degree of $Q(x)$, there is no horizontal asymptote.

$$\text{Consider } g(x) = \frac{2x^2 - 3}{3x^2 + 5}.$$

Here, the degrees of numerator and denominator are equal, thus the horizontal asymptote is $y = \frac{2}{3}$.

Finding Intercepts

The x-intercepts are found by setting $y = 0$, which requires solving $P(x) = 0$, since any value of x making $P(x) = 0$ will make the whole function zero. The y-intercept is found by evaluating $f(0)$.

$$\text{For } f(x) = \frac{x - 2}{x^2 - 4},$$

x-intercept is at $x = 2$ (since $x - 2 = 0$ when $x = 2$), and the y-intercept is $f(0) = \frac{0-2}{0^2-4} = \frac{1}{2}$.

Plotting Points

After determining the horizontal and vertical asymptotes and intercepts, plot a few more points to get a clearer picture of the graph. Choose values of x around the intercepts and asymptotes, and calculate corresponding y values.

Sketching the Graph

With the asymptotes, intercepts, and additional points identified, sketch the graph. The graph will approach the asymptotes without crossing them. Between vertical asymptotes, the graph can take various shapes but remember, it cannot cross a vertical asymptote.

Identifying Discontinuities

Discontinuities in a rational function occur at values of x that make the denominator equal to zero, leading to undefined points unless there's a common factor in both the numerator and denominator that cancels out. Plotting any removable discontinuities as holes on the graph is an essential part of graphing rational functions.

Behavior Near Asymptotes

The graph's behavior near the asymptotes can be determined by analyzing the sign of $f(x)$ on either side of the vertical asymptotes and as x approaches $\pm\infty$ for horizontal asymptotes. These analyses give insight into whether the graph approaches the asymptotes from above or below.

Graphing rational functions requires a systematic approach that includes identifying asymptotes and intercepts, plotting key points, and sketching the function's general shape while paying attention to its behavior near the asymptotes and discontinuities. With practice, the process becomes intuitive, allowing for the accurate graphing of complex rational functions.

10.10 Asymptotes and Discontinuities

Rational functions, which are ratios of polynomials, exhibit unique characteristics in their graphs. Among these are asymptotes and discontinuities, which are essential in understanding the behavior of these functions. An asymptote is a line that the graph of a function approaches but never touches. Discontinuities, on the other hand, are points or intervals where a function is undefined or its limit does not exist.

Vertical Asymptotes

Vertical asymptotes occur at values of x where the denominator of a rational function equals zero, and the numerator does not equal zero at the same point. To find vertical asymptotes, one must first factor the numerator and the denominator of the function, if possible. Setting the denominator equal to zero and solving for x will yield the potential vertical asymptotes. It's important to check each solution to ensure it does not make the numerator zero as well, as this may indicate the presence of a hole rather than an asymptote.

10.10. ASYMPTOTES AND DISCONTINUITIES

$$\text{Given } f(x) = \frac{x^2 - 4}{x^2 - 5x + 6},$$

set denominator $x^2 - 5x + 6 = 0$.

Solving this, we find $x = 2$ or $x = 3$. These are the points where $f(x)$ might have vertical asymptotes.

Horizontal Asymptotes

Horizontal asymptotes are approached by the graph of the function as x approaches infinity or negative infinity. The presence and location of horizontal asymptotes depend on the degrees of the numerator and denominator polynomials.

- If the degree of the numerator is less than the degree of the denominator, the horizontal asymptote is $y = 0$.

- If the degrees are equal, the horizontal asymptote is $y = \frac{a}{b}$, where a and b are the leading coefficients of the numerator and denominator, respectively.

- If the degree of the numerator is greater than the degree of the denominator, there is no horizontal asymptote.

Slant Asymptotes

Slant or oblique asymptotes occur when the degree of the numerator is exactly one greater than the degree of the denominator. To find the slant asymptote, perform polynomial long division or synthetic division of the numerator by the denominator. The quotient (excluding the remainder) represents the equation of the slant asymptote.

$$\text{For } f(x) = \frac{2x^2 - 3x - 5}{x - 2},$$
long division yields $y = 2x + 1$.

Here, $y = 2x + 1$ is the slant asymptote.

Discontinuities

Discontinuities in rational functions arise when the function is undefined, which occurs at points where the denominator is zero. These can be classified into removable discontinuities, where the function can be redefined to make it continuous, and non-removable discontinuities, which include jumps or vertical asymptotes.

Identifying Discontinuities

Setting the denominator equal to zero as with vertical asymptotes, identify the x values. If a factor in the denominator also zeros out the numerator, the discontinuity may be removable if the factor can be canceled out.

$$\text{Consider } f(x) = \frac{x^2 - 4}{x^2 - 4x + 4},$$
$$(x^2 - 4x + 4) = 0 \implies (x - 2)^2 = 0.$$

Here, $x = 2$ creates a discontinuity. Since the same factor does not zero out the numerator, this discontinuity is non-removable, indicating a vertical asymptote.

10.10. ASYMPTOTES AND DISCONTINUITIES

Graphical Representation

Understanding asymptotes and discontinuities is crucial for sketching the graph of rational functions. Vertical and horizontal asymptotes provide a framework, indicating where the function approaches but does not touch or cross these lines. Slant asymptotes offer a similar guideline for the end behavior of the function for large magnitudes of x. Recognizing discontinuities allows for the identification of 'holes' in the graph. When graphing, plot these features first to guide the sketching of the rational function.

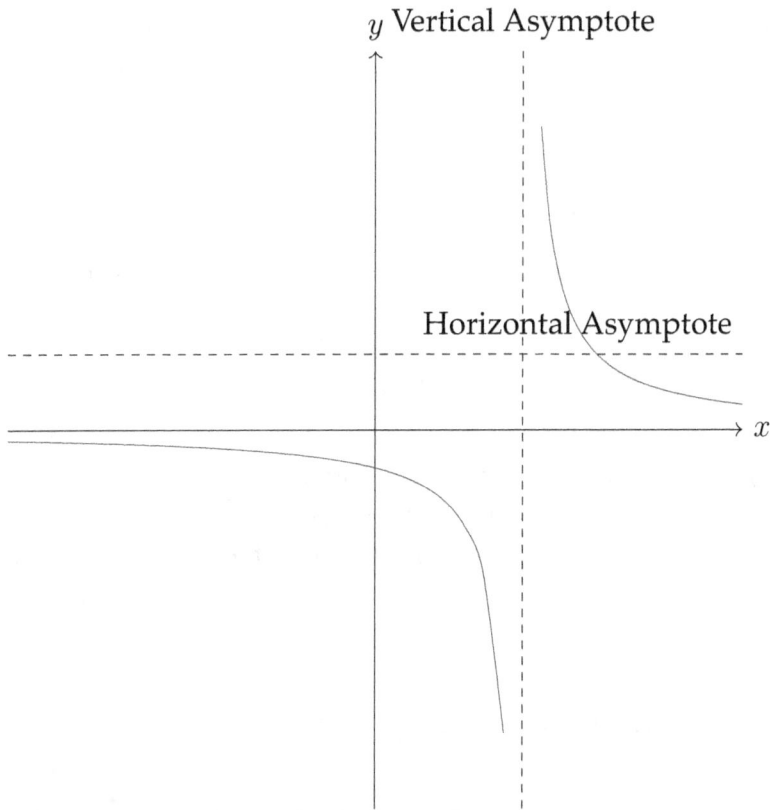

In summary, rational functions often exhibit asymptotes

345

and discontinuities which impact their graphical representations. By understanding these concepts and how to identify them, one can accurately sketch and interpret the behavior of rational functions.

10.11 Inverse Variation

In the study of rational expressions, the concept of inverse variation plays a crucial role in modeling relationships where one quantity varies inversely as another. Unlike direct variation, where the two quantities increase or decrease together, in inverse variation, an increase in one quantity results in a corresponding decrease in the other, and vice versa. This section dives into the mathematical representation of inverse variation, typical applications, and methods of solving problems involving this concept.

An inverse variation can be represented by the equation:

$$y = \frac{k}{x}$$

where y varies inversely as x, and k is a nonzero constant known as the constant of variation. The value of k can be found if a pair of quantities x and y that follow the inverse variation trend are known.

Examples of Inverse Variation

For instance, if it is given that y varies inversely as x, and $y = 2$ when $x = 3$, one can determine the constant of

10.11. INVERSE VARIATION

variation using the formula:

$$2 = \frac{k}{3}$$
$$k = 6$$

This implies that the equation governing the relationship between x and y is $y = \frac{6}{x}$.

Properties of Inverse Variation

Key properties of inverse variations include:

- The product xy is constant for all values of x and y that adhere to the inverse variation.

- The graph of an inverse variation equation is a hyperbola.

- As x approaches infinity, y approaches zero, indicating an asymptotic behavior towards the x-axis.

- Similarly, as x approaches zero, y approaches infinity, indicating an asymptotic behavior towards the y-axis.

Graphing Inverse Variations

To graph an equation of the form $y = \frac{k}{x}$, one must:

- Plot enough points (x, y) that satisfy the equation to reveal the curve's shape. Typically, values of x that are both positive and negative are chosen.

- Draw the curve that connects the plotted points, keeping in mind the asymptotic behavior towards both axes.

- Label the curve with its equation for clarity.

These steps can be demonstrated using the equation $y = \frac{6}{x}$.

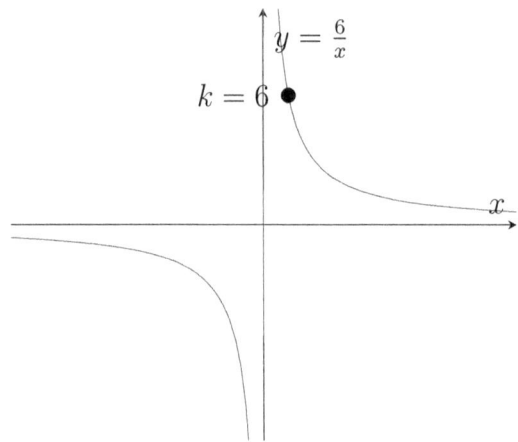

Applications of Inverse Variation

Inverse variations are commonly seen in physics, economics, and other fields. For example, in physics, the intensity of light or sound waves inversely varies with the square of the distance from the source. In economics, the concept is applied in the demand and supply model, where the price of a good might inversely vary with the quantity available.

10.12. APPLICATIONS OF RATIONAL EQUATIONS AND MODELS

Solving Problems Involving Inverse Variation

Problem-solving with inverse variation generally involves identifying the constant of variation k and using it to find unknown values of x or y given the other. It's essential to recognize the problem's context to apply the inverse variation principle correctly.

For instance, if a car travels at constant speed, the time taken (t) to cover a distance (d) varies inversely with the speed (v) of the car. The relationship can be represented as $t = \frac{k}{v}$, where k is the constant of variation, which, in this context, represents the distance.

By grasping the inverse variation principle and its applications, one can resolve a wide array of problems in mathematics and beyond, illustrating the power and versatility of rational expressions in modeling real-world phenomena.

10.12 Applications of Rational Equations and Models

Rational equations, which involve ratios of polynomials, find extensive applicability in various fields including physics, engineering, economics, and biology. This section delves into the practical applications of rational equations and models, elucidating the theoretical knowledge acquired in previous sections through real-world scenarios. A rational equation is defined as an equation in which two rational expressions are set equal to each other. These equations are not only ubiquitous in mathematical problems but also in everyday life situations.

One common application of rational equations is in the domain of speed, distance, and time problems. In such problems, the relationship between speed (v), distance (d), and time (t) is given by the equation $v = \frac{d}{t}$. By manipulating this equation, one can model situations to solve for any of the three variables given the other two.

$$t = \frac{d}{v}$$

Consider a scenario where two cities are 300 miles apart, and a car travels from one city to the other at an average speed that can be represented as a rational expression. This situation can be modeled with rational equations to find the time taken for the journey or the average speed, given the time.

Another pivotal application of rational equations is in work rate problems, which are encountered in both academic and professional settings. If an individual or a machine can complete a certain job in x hours, then the work rate is given by $\frac{1}{x}$. Work rates can additively combine in scenarios where multiple entities are working together towards a common task.

$$\text{Combined work rate} = \frac{1}{x} + \frac{1}{y}$$

For example, if two pipes are filling a pool and one can fill the pool in 2 hours while the other in 3 hours, the time it takes for both pipes to fill the pool together can be found by setting up a rational equation involving their combined work rates.

Rational equations are also fundamental in the formulation of concentration problems in chemistry. The concentration of solutions, the mixture of different solutions to

10.12. APPLICATIONS OF RATIONAL EQUATIONS AND MODELS

achieve a desired concentration, and the dilution of solutions can all be modeled using rational equations. The mixture problem, in particular, involves adding two solutions of different concentrations to get a new solution of a desired concentration.

$$\text{Amount of Solution 1} \times \text{Concentration of Solution 1} +$$
$$\text{Amount of Solution 2} \times \text{Concentration of Solution 2}$$
$$= \text{Total Amount} \times \text{Desired Concentration}$$

Consider mixing a 20% salt solution with a 50% salt solution to get 10 liters of a 30% salt solution. We can establish a rational equation based on the quantities and concentrations of the solutions involved.

Economics and business applications often involve cost, revenue, and profit functions, which can be represented as rational expressions. These functions allow for the analysis of break-even points, optimization of profit, and understanding of cost behaviors.

$$\text{Profit} = \text{Revenue} - \text{Cost}$$

If the revenue and cost are given as rational expressions of the quantity of goods sold or produced, finding the break-even point or maximizing profit involves solving rational equations.

Example: Let Revenue = $\frac{200x}{1+2x}$ and Cost = $\frac{150x}{2+x}$. To find the break-even point, set Revenue = Cost and solve for x.

Engineering often utilizes rational models to describe relationships between variables in system designs and analysis, such as in the flow rates of fluids through systems, electrical circuits, and mechanical systems' efficiencies.

In summary, rational equations and models are indispensable tools in modeling and solving real-world problems across various domains. These applications underscore the importance of understanding and effectively utilizing rational equations to interpret and navigate the complexities of everyday situations and professional challenges. Through the discussed examples, we have illustrated the practical significance of rational expressions and equations, bridging the gap between theoretical mathematics and its real-world applications.

www.ingramcontent.com/pod-product-compliance
Lightning Source LLC
Chambersburg PA
CBHW052139220526
45471CB00004B/1445